T0350163

An Elementary Introduction to Mathematical Finance, Third Edition

This textbook on the basics of option pricing is accessible to readers with limited mathematical training. It is for both professional traders and undergraduates studying the basics of finance. Assuming no prior knowledge of probability, Sheldon M. Ross offers clear, simple explanations of arbitrage, the Black–Scholes option pricing formula, and other topics such as utility functions, optimal portfolio selections, and the capital assets pricing model. Among the many new features of this third edition are new chapters on Brownian motion and geometric Brownian motion, stochastic order relations, and stochastic dynamic programming, along with expanded sets of exercises and references for all the chapters.

Sheldon M. Ross is the Epstein Chair Professor in the Department of Industrial and Systems Engineering, University of Southern California. He received his Ph.D. in statistics from Stanford University in 1968 and was a Professor at the University of California, Berkeley, from 1976 until 2004. He has published more than 100 articles and a variety of textbooks in the areas of statistics and applied probability, including *Topics in Finite and Discrete Mathematics* (2000), *Introduction to Probability and Statistics for Engineers and Scientists, Fourth Edition* (2009), *A First Course in Probability, Eighth Edition* (2009), and *Introduction to Probability Models, Tenth Edition* (2009). Dr. Ross serves as the editor for *Probability in the Engineering and Informational Sciences*.

An Elementary Introduction to Mathematical Finance

Third Edition

SHELDON M. ROSS
University of Southern California

CAMBRIDGE
UNIVERSITY PRESS

CAMBRIDGE UNIVERSITY PRESS
Cambridge, New York, Melbourne, Madrid, Cape Town,
Singapore, São Paulo, Delhi, Mexico City

Cambridge University Press
32 Avenue of the Americas, New York, NY 10013-2473, USA

www.cambridge.org
Information on this title: www.cambridge.org/9780521192538

First published 1999
Second edition published 2003
Third edition published 2011
Reprinted 2011, 2012

A catalog record for this publication is available from the British Library.

Library of Congress Cataloging in Publication Data

Ross, Sheldon M. (Sheldon Mark), 1943–
An elementary introduction to mathematical finance / Sheldon M. Ross. – Third edition.
  p. cm.
Includes index.
ISBN 978-0-521-19253-8
1. Investments – Mathematics. 2. Stochastic analysis. 3. Options (Finance) –
Mathematical models. 4. Securities – Prices – Mathematical models. I. Title.
HG4515.3.R67 2011
332.601'51–dc22   2010049863

ISBN 978-0-521-19253-8 Hardback

To my parents,
Ethel and Louis Ross

Contents

Introduction and Preface

An *option* gives one the right, but not the obligation, to buy or sell a security under specified terms. A *call option* is one that gives the right to buy, and a *put option* is one that gives the right to sell the security. Both types of options will have an *exercise price* and an *exercise time*. In addition, there are two standard conditions under which options operate: *European* options can be utilized only at the exercise time, whereas *American* options can be utilized at any time up to exercise time. Thus, for instance, a European call option with exercise price K and exercise time t gives its holder the right to purchase at time t one share of the underlying security for the price K, whereas an American call option gives its holder the right to make the purchase at any time before or at time t.

A prerequisite for a strong market in options is a computationally efficient way of evaluating, at least approximately, their worth; this was accomplished for call options (of either American or European type) by the famous Black–Scholes formula. The formula assumes that prices of the underlying security follow a geometric Brownian motion. This means that if $S(y)$ is the price of the security at time y then, for any price history up to time y, the ratio of the price at a specified future time $t + y$ to the price at time y has a lognormal distribution with mean and variance parameters $t\mu$ and $t\sigma^2$, respectively. That is,

$$\log\left(\frac{S(t + y)}{S(y)}\right)$$

will be a normal random variable with mean $t\mu$ and variance $t\sigma^2$. Black and Scholes showed, under the assumption that the prices follow a geometric Brownian motion, that there is a single price for a call option that does not allow an idealized trader – one who can instantaneously make trades without any transaction costs – to follow a strategy that will result in a sure profit in all cases. That is, there will be no certain profit (i.e., no *arbitrage*) if and only if the price of the option is as given by the Black–Scholes formula. In addition, this price depends only on the variance parameter σ of the geometric Brownian motion (as well as on the prevailing interest rate, the underlying price of the security, and the conditions of the option) and not on the parameter μ. Because the parameter σ is a measure of the volatility of the security, it is often called the *volatility* parameter.

A *risk-neutral* investor is one who values an investment solely through the expected present value of its return. If such an investor models a security by a geometric Brownian motion that turns all investments involving buying and selling the security into fair bets, then this investor's valuation of a call option on this security will be precisely as given by the Black–Scholes formula. For this reason, the Black–Scholes valuation is often called a *risk-neutral valuation.*

Our first objective in this book is to derive and explain the Black–Scholes formula. Its derivation, however, requires some knowledge of probability, and this is what the first three chapters are concerned with. Chapter 1 introduces probability and the probability experiment. Random variables – numerical quantities whose values are determined by the outcome of the probability experiment – are discussed, as are the concepts of the expected value and variance of a random variable. In Chapter 2 we introduce normal random variables; these are random variables whose probabilities are determined by a bell-shaped curve. The central limit theorem is presented in this chapter. This theorem, probably the most important theoretical result in probability, states that the sum of a large number of random variables will approximately be a normal random variable. In Chapter 3 we introduce the geometric Brownian motion process; we define it, show how it can be obtained as the limit of simpler processes, and discuss the justification for its use in modeling security prices.

With the probability necessities behind us, the second part of the text begins in Chapter 4 with an introduction to the concept of interest rates and present values. A key concept underlying the Black–Scholes formula is that of arbitrage, which is the subject of Chapter 5. In this chapter we show how arbitrage can be used to determine prices in a variety of situations, including the single-period binomial option model. In Chapter 6 we present the arbitrage theorem and use it to find an expression for the unique nonarbitrage option cost in the multiperiod binomial model. In Chapter 7 we use the results of Chapter 6, along with the approximations of geometric Brownian motion presented in Chapter 4, to obtain a simple derivation of the Black–Scholes equation for pricing call options. Properties of the resultant option cost as a function of its parameters are derived, as is the delta hedging replication strategy. Additional results on options are presented in Chapter 8, where we derive option prices for dividend-paying securities; show how to utilize a multiperiod binomial model to determine an approximation of the risk-neutral price of an American put option; determine no-arbitrage costs when the security's price follows a model that superimposes random jumps on a geometric Brownian motion; and present different estimators of the volatility parameter.

In Chapter 9 we note that, in many situations, arbitrage considerations do not result in a unique cost. We show the importance in such cases of the investor's utility function as well as his or her estimates of the probabilities of the possible outcomes of the investment. The concepts of mean variance analysis, value and conditional value at risk, and the capital assets pricing model are introduced.

In Chapter 10 we introduce stochastic order relations. These relations can be useful in determining which of a class of investments is best without completely specifying the investor's utility function. For instance, if the return from one investment is greater than the return from another investment in the sense of first-order stochastic dominance, then the first investment is to be preferred for any increasing utility function; whereas if the first return is greater in the sense of second-order stochastic dominance, then the first investment is to be preferred as long as the utility function is concave and increasing.

In Chapters 11 and 12 we study some optimization models in finance. In Chapter 13 we introduce some nonstandard, or "exotic," options such as barrier, Asian, and lookback options. We explain how to use Monte Carlo simulation, implementing variance reduction techniques, to efficiently determine their geometric Brownian motion risk-neutral valuations.

The Black–Scholes formula is useful even if one has doubts about the validity of the underlying geometric Brownian model. For as long as one accepts that this model is at least approximately valid, its use gives one an idea about the *appropriate* price of the option. Thus, if the actual trading option price is below the formula price then it would seem that the option is underpriced in relation to the security itself, thus leading one to consider a strategy of buying options and selling the security (with the reverse being suggested when the trading option price is above the formula price). In Chapter 14 we show that real data cannot aways be fit by a geometric Brownian motion model, and that more general models may need to be considered. In the case of commodity prices, there is a strong belief by many traders in the concept of mean price reversion: that the market prices of certain commodities have tendencies to revert to fixed values. In Chapter 15 we present a model, more general than geometric Brownian motion, that can be used to model the price flow of such a commodity.

New to This Edition

Whereas the third edition contains changes in almost all previous chapters, the major changes in the new edition are as follows.

• Chapter 3 on *Brownian Motion and Geometric Brownian Motion* has been completely rewritten. Among other things the new chapter gives an elementary derivation of the distribution of the maximum variable of a Brownian motion process with drift, as well as an elementary proof of the Cameron–Martin theorem.
• Section 7.5.2 has been reworked, clarifying the argument leading to a simple derivation of the partial derivatives of the Black–Scholes call option pricing formula.
• Section 7.6 on *European Put Options* is new. It presents monotonicity and convexity results concerning the risk-neutral price of a European put option.
• Chapter 10 on *Stochastic Order Relations* is new. This chapter presents first- and second-order stochastic dominance, as well as likelihood ratio orderings. Among other things, it is shown (in Section 10.5.1) that a normal random variable decreases, in the second-order stochastic dominance sense, as its variance increases.
• The old Chapter 10 is now Chapter 11.
• Chapter 12 on *Stochastic Dynamic Programming* is new.
• The old Chapter 11 is now Chapter 13. New within this chapter is Section 13.9, which presents continuous time approximations of barrier and lookback options.
• The old Chapter 12 is now Chapter 14.
• The old Chapter 13 is now Chapter 15.

One technical point that should be mentioned is that we use the notation $\log(x)$ to represent the natural logarithm of x. That is, the logarithm has base e, where e is defined by

$$e = \lim_{n \to \infty} (1 + 1/n)^n$$

and is approximately given by $2.71828\ldots$.

We would like to thank Professors Ilan Adler and Shmuel Oren for some enlightening conversations, Mr. Kyle Lin for his many useful comments, and Mr. Nahoya Takezawa for his general comments and for doing the numerical work needed in the final chapters. We would also like to thank Professors Anthony Quas, Daniel Naiman, and Agostino Capponi for helpful comments concerning the previous edition.

1. Probability

1.1 Probabilities and Events

Consider an experiment and let S, called the *sample space*, be the set of all possible outcomes of the experiment. If there are m possible outcomes of the experiment then we will generally number them 1 through m, and so $S = \{1, 2, \ldots, m\}$. However, when dealing with specific examples, we will usually give more descriptive names to the outcomes.

Example 1.1a (i) Let the experiment consist of flipping a coin, and let the outcome be the side that lands face up. Thus, the sample space of this experiment is

$$S = \{h, t\},$$

where the outcome is h if the coin shows heads and t if it shows tails.

(ii) If the experiment consists of rolling a pair of dice – with the outcome being the pair (i, j), where i is the value that appears on the first die and j the value on the second – then the sample space consists of the following 36 outcomes:

$$(1, 1), \ (1, 2), \ (1, 3), \ (1, 4), \ (1, 5), \ (1, 6),$$

$$(2, 1), \ (2, 2), \ (2, 3), \ (2, 4), \ (2, 5), \ (2, 6),$$

$$(3, 1), \ (3, 2), \ (3, 3), \ (3, 4), \ (3, 5), \ (3, 6),$$

$$(4, 1), \ (4, 2), \ (4, 3), \ (4, 4), \ (4, 5), \ (4, 6),$$

$$(5, 1), \ (5, 2), \ (5, 3), \ (5, 4), \ (5, 5), \ (5, 6),$$

$$(6, 1), \ (6, 2), \ (6, 3), \ (6, 4), \ (6, 5), \ (6, 6).$$

(iii) If the experiment consists of a race of r horses numbered 1, 2, 3, $\ldots, r$, and the outcome is the order of finish of these horses, then the sample space is

$$S = \{\text{all orderings of the numbers } 1, 2, 3, \ldots, r\}.$$

For instance, if $r = 4$ then the outcome is $(1, 4, 2, 3)$ if the number 1 horse comes in first, number 4 comes in second, number 2 comes in third, and number 3 comes in fourth. □

Consider once again an experiment with the sample space $S = \{1, 2, \ldots, m\}$. We will now suppose that there are numbers $p_1, \ldots, p_m$ with

$$p_i \geq 0, \ i = 1, \ldots, m, \quad \text{and} \quad \sum_{i=1}^{m} p_i = 1$$

and such that p_i is the *probability* that i is the outcome of the experiment.

Example 1.1b In Example 1.1a(i), the coin is said to be *fair* or *unbiased* if it is equally likely to land on heads as on tails. Thus, for a fair coin we would have that

$$p_h = p_t = 1/2.$$

If the coin were biased and heads were twice as likely to appear as tails, then we would have

$$p_h = 2/3, \qquad p_t = 1/3.$$

If an unbiased pair of dice were rolled in Example 1.1a(ii), then all possible outcomes would be equally likely and so

$$p_{(i,j)} = 1/36, \ \ 1 \leq i \leq 6, \ 1 \leq j \leq 6.$$

If $r = 3$ in Example 1.1a(iii), then we suppose that we are given the six nonnegative numbers that sum to 1:

$$p_{1,2,3}, \ p_{1,3,2}, \ p_{2,1,3}, \ p_{2,3,1}, \ p_{3,1,2}, \ p_{3,2,1},$$

where $p_{i,j,k}$ represents the probability that horse i comes in first, horse j second, and horse k third. □

Any set of possible outcomes of the experiment is called an *event*. That is, an event is a subset of S, the set of all possible outcomes. For any event A, we say that A *occurs* whenever the outcome of the experiment is a point in A. If we let $P(A)$ denote the probability that event A occurs, then we can determine it by using the equation

$$P(A) = \sum_{i \in A} p_i. \tag{1.1}$$

Note that this implies

$$P(S) = \sum_i p_i = 1. \tag{1.2}$$

In words, the probability that the outcome of the experiment is in the sample space is equal to 1 – which, since S consists of all possible outcomes of the experiment, is the desired result.

Example 1.1c   Suppose the experiment consists of rolling a pair of fair dice. If A is the event that the sum of the dice is equal to 7, then

$$A = \{(1, 6), (2, 5), (3, 4), (4, 3), (5, 2), (6, 1)\}$$

and

$$P(A) = 6/36 = 1/6.$$

If we let B be the event that the sum is 8, then

$$P(B) = p_{(2,6)} + p_{(3,5)} + p_{(4,4)} + p_{(5,3)} + p_{(6,2)} = 5/36.$$

If, in a horse race between three horses, we let A denote the event that horse number 1 wins, then $A = \{(1, 2, 3), (1, 3, 2)\}$ and

$$P(A) = p_{1,2,3} + p_{1,3,2}. \qquad \square$$

For any event A, we let A^c, called the *complement* of A, be the event containing all those outcomes in S that are not in A. That is, A^c occurs if and only if A does not. Since

$$1 = \sum_i p_i$$

$$= \sum_{i \in A} p_i + \sum_{i \in A^c} p_i$$

$$= P(A) + P(A^c),$$

we see that

$$P(A^c) = 1 - P(A). \tag{1.3}$$

That is, the probability that the outcome is not in A is 1 minus the probability that it is in A. The complement of the sample space S is the null event $\emptyset$, which contains no outcomes. Since $\emptyset = S^c$, we obtain from

Equations (1.2) and (1.3) that

$$\dot{P}(\emptyset) = 0.$$

For any events A and B we define $A \cup B$, called the *union* of A and B, as the event consisting of all outcomes that are in A, or in B, or in both A and B. Also, we define their *intersection* AB (sometimes written $A \cap B$) as the event consisting of all outcomes that are both in A and in B.

Example 1.1d Let the experiment consist of rolling a pair of dice. If A is the event that the sum is 10 and B is the event that both dice land on even numbers greater than 3, then

$$A = \{(4, 6), (5, 5), (6, 4)\}, \qquad B = \{(4, 4), (4, 6), (6, 4), (6, 6)\}.$$

Therefore,

$$A \cup B = \{(4, 4), (4, 6), (5, 5), (6, 4), (6, 6)\},$$

$$AB = \{(4, 6), (6, 4)\}. \qquad \qquad \square$$

For any events A and B, we can write

$$P(A \cup B) = \sum_{i \in A \cup B} p_i,$$

$$P(A) = \sum_{i \in A} p_i,$$

$$P(B) = \sum_{i \in B} p_i.$$

Since every outcome in both A and B is counted twice in $P(A) + P(B)$ and only once in $P(A \cup B)$, we obtain the following result, often called the *addition theorem of probability*.

Proposition 1.1.1

$$P(A \cup B) = P(A) + P(B) - P(AB).$$

Thus, the probability that the outcome of the experiment is either in A or in B equals the probability that it is in A, plus the probability that it is in B, minus the probability that it is in both A and B.

Example 1.1e Suppose the probabilities that the Dow-Jones stock index increases today is .54, that it increases tomorrow is .54, and that it increases both days is .28. What is the probability that it does not increase on either day?

Solution. Let A be the event that the index increases today, and let B be the event that it increases tomorrow. Then the probability that it increases on at least one of these days is

$$P(A \cup B) = P(A) + P(B) - P(AB)$$

$$= .54 + .54 - .28 = .80.$$

Therefore, the probability that it increases on neither day is $1 - .80 = .20$.   □

If $AB = \emptyset$, we say that A and B are *mutually exclusive* or *disjoint*. That is, events are mutually exclusive if they cannot both occur. Since $P(\emptyset) = 0$, it follows from Proposition 1.1.1 that, when A and B are mutually exclusive,

$$P(A \cup B) = P(A) + P(B).$$

1.2   Conditional Probability

Suppose that each of two teams is to produce an item, and that the two items produced will be rated as either acceptable or unacceptable. The sample space of this experiment will then consist of the following four outcomes:

$$S = \{(a, a), (a, u), (u, a), (u, u)\},$$

where (a, u) means, for instance, that the first team produced an acceptable item and the second team an unacceptable one. Suppose that the probabilities of these outcomes are as follows:

$$P(a, a) = .54,$$

$$P(a, u) = .28,$$

$$P(u, a) = .14,$$

$$P(u, u) = .04.$$

If we are given the information that exactly one of the items produced was acceptable, what is the probability that it was the one produced by the first team? To determine this probability, consider the following reasoning. Given that there was exactly one acceptable item produced, it follows that the outcome of the experiment was either (a, u) or (u, a). Since the outcome (a, u) was initially twice as likely as the outcome (u, a), it should remain twice as likely given the information that one of them occurred. Therefore, the probability that the outcome was (a, u) is $2/3$, whereas the probability that it was (u, a) is $1/3$.

Let $A = \{(a, u), (a, a)\}$ denote the event that the item produced by the first team is acceptable, and let $B = \{(a, u), (u, a)\}$ be the event that exactly one of the produced items is acceptable. The probability that the item produced by the first team was acceptable given that exactly one of the produced items was acceptable is called the *conditional probability* of A given that B has occurred; this is denoted as

$$P(A|B).$$

A general formula for $P(A|B)$ is obtained by an argument similar to the one given in the preceding. Namely, if the event B occurs then, in order for the event A to occur, it is necessary that the occurrence be a point in both A and B; that is, it must be in AB. Now, since we know that B has occurred, it follows that B can be thought of as the new sample space, and hence the probability that the event AB occurs will equal the probability of AB relative to the probability of B. That is,

$$P(A|B) = \frac{P(AB)}{P(B)}. \tag{1.4}$$

Example 1.2a   A coin is flipped twice. Assuming that all four points in the sample space $S = \{(h, h), (h, t), (t, h), (t, t)\}$ are equally likely, what is the conditional probability that both flips land on heads, given that

(a) the first flip lands on heads, and
(b) at least one of the flips lands on heads?

Solution.   Let $A = \{(h, h)\}$ be the event that both flips land on heads; let $B = \{(h, h), (h, t)\}$ be the event that the first flip lands on heads; and let $C = \{(h, h), (h, t), (t, h)\}$ be the event that at least one of the flips lands on heads. We have the following solutions:

$$P(A|B) = \frac{P(AB)}{P(B)}$$

$$= \frac{P(\{(h, h)\})}{P(\{(h, h), (h, t)\})}$$

$$= \frac{1/4}{2/4}$$

$$= 1/2$$

and

$$P(A|C) = \frac{P(AC)}{P(C)}$$

$$= \frac{P(\{(h, h)\})}{P(\{(h, h), (h, t), (t, h)\})}$$

$$= \frac{1/4}{3/4}$$

$$= 1/3.$$

Many people are initially surprised that the answers to parts (a) and (b) are not identical. To understand why the answers are different, note first that – conditional on the first flip landing on heads – the second one is still equally likely to land on either heads or tails, and so the probability in part (a) is $1/2$. On the other hand, knowing that at least one of the flips lands on heads is equivalent to knowing that the outcome is not (t, t). Thus, given that at least one of the flips lands on heads, there remain three equally likely possibilities, namely $(h, h), (h, t), (t, h)$, showing that the answer to part (b) is $1/3$. □

It follows from Equation (1.4) that

$$P(AB) = P(B)P(A|B). \tag{1.5}$$

That is, the probability that both A and B occur is the probability that B occurs multiplied by the conditional probability that A occurs given that B occurred; this result is often called the *multiplication theorem of probability*.

Example 1.2b Suppose that two balls are to be withdrawn, without replacement, from an urn that contains 9 blue and 7 yellow balls. If each ball drawn is equally likely to be any of the balls in the urn at the time, what is the probability that both balls are blue?

Solution. Let B_1 and B_2 denote, respectively, the events that the first and second balls withdrawn are blue. Now, given that the first ball withdrawn is blue, the second ball is equally likely to be any of the remaining 15 balls, of which 8 are blue. Therefore, $P(B_2|B_1) = 8/15$. As $P(B_1) = 9/16$, we see that

$$P(B_1 B_2) = \frac{9}{16}\frac{8}{15} = \frac{3}{10}.$$ □

The conditional probability of A given that B has occurred is not generally equal to the unconditional probability of A. In other words, knowing that the outcome of the experment is an element of B generally changes the probability that it is an element of A. (What if A and B are mutually exclusive?) In the special case where $P(A|B)$ is equal to $P(A)$, we say that A is *independent* of B. Since

$$P(A|B) = \frac{P(AB)}{P(B)},$$

we see that A is independent of B if

$$P(AB) = P(A)P(B).   (1.6)$$

The relation in (1.6) is symmetric in A and B. Thus it follows that, whenever A is independent of B, B is also independent of A – that is, A and B are *independent events*.

Example 1.2c Suppose that, with probability .52, the closing price of a stock is at least as high as the close on the previous day, and that the results for succesive days are independent. Find the probability that the closing price goes down in each of the next four days, but not on the following day.

Solution. Let A_i be the event that the closing price goes down on day i. Then, by independence, we have

$$P(A_1 A_2 A_3 A_4 A_5^c) = P(A_1)P(A_2)P(A_3)P(A_4)P(A_5^c)$$

$$= (.48)^4(.52) = .0276.$$ □

1.3   Random Variables and Expected Values

Numerical quantities whose values are determined by the outcome of the experiment are known as *random variables*. For instance, the sum obtained when rolling dice, or the number of heads that result in a series of coin flips, are random variables. Since the value of a random variable is determined by the outcome of the experiment, we can assign probabilities to each of its possible values.

Example 1.3a   Let the random variable X denote the sum when a pair of fair dice are rolled. The possible values of X are 2, 3, ..., 12, and they have the following probabilities:

$$P\{X = 2\} = P\{(1, 1)\} = 1/36,$$

$$P\{X = 3\} = P\{(1, 2), (2, 1)\} = 2/36,$$

$$P\{X = 4\} = P\{(1, 3), (2, 2), (3, 1)\} = 3/36,$$

$$P\{X = 5\} = P\{(1, 4), (2, 3), (3, 2), (4, 1)\} = 4/36,$$

$$P\{X = 6\} = P\{(1, 5), (2, 4), (3, 3), (4, 2), (5, 1)\} = 5/36,$$

$$P\{X = 7\} = P\{(1, 6), (2, 5), (3, 4), (4, 3), (5, 2), (6, 1)\} = 6/36,$$

$$P\{X = 8\} = P\{(2, 6), (3, 5), (4, 4), (5, 3), (6, 2)\} = 5/36,$$

$$P\{X = 9\} = P\{(3, 6), (4, 5), (5, 4), (6, 3)\} = 4/36,$$

$$P\{X = 10\} = P\{(4, 6), (5, 5), (6, 4)\} = 3/36,$$

$$P\{X = 11\} = P\{(5, 6), (6, 5)\} = 2/36,$$

$$P\{X = 12\} = P\{(6, 6)\} = 1/36. \qquad \square$$

If X is a random variable whose possible values are $x_1, x_2, \ldots, x_n$, then the set of probabilities $P\{X = x_j\}$ ($j = 1, \ldots, n$) is called the *probability distribution* of the random variable. Since X must assume one of these values, it follows that

$$\sum_{j=1}^{n} P\{X = x_j\} = 1.$$

Definition   If X is a random variable whose possible values are $x_1, x_2, \ldots, x_n$, then the *expected value* of X, denoted by $E[X]$, is defined by

$$E[X] = \sum_{j=1}^{n} x_j P\{X = x_j\}.$$

Alternative names for $E[X]$ are the *expectation* or the *mean* of X.

In words, $E[X]$ is a weighted average of the possible values of X, where the weight given to a value is equal to the probability that X assumes that value.

Example 1.3b   Let the random variable X denote the amount that we win when we make a certain bet. Find $E[X]$ if there is a 60% chance that we lose 1, a 20% chance that we win 1, and a 20% chance that we win 2.

Solution.
$$E[X] = -1(.6) + 1(.2) + 2(.2) = 0.$$

Thus, the expected amount that is won on this bet is equal to 0. A bet whose expected winnings is equal to 0 is called a *fair* bet.   □

Example 1.3c   A random variable X, which is equal to 1 with probability p and to 0 with probability $1 - p$, is said to be a *Bernoulli* random variable with parameter p. Its expected value is

$$E[X] = 1(p) + 0(1 - p) = p.$$ □

A useful and easily established result is that, for constants a and b,

$$E[aX + b] = aE[X] + b. \tag{1.7}$$

To verify Equation (1.7), let $Y = aX + b$. Since Y will equal $ax_j + b$ when $X = x_j$, it follows that

$$E[Y] = \sum_{j=1}^{n} (ax_j + b) P\{X = x_j\}$$

$$= \sum_{j=1}^{n} ax_j P\{X = x_j\} + \sum_{j=1}^{n} bP\{X = x_j\}$$

$$= a \sum_{j=1}^{n} x_j P\{X = x_j\} + b \sum_{j=1}^{n} P\{X = x_j\}$$

$$= aE[X] + b.$$

An important result is that the expected value of a sum of random variables is equal to the sum of their expected values.

Proposition 1.3.1 *For random variables* $X_1, \ldots, X_k$,

$$E\left[\sum_{j=1}^{k} X_j\right] = \sum_{j=1}^{k} E[X_j].$$

Example 1.3d Consider n independent trials, each of which is a success with probability p. The random variable X, equal to the total number of successes that occur, is called a *binomial* random variable with parameters n and p. To determine the probability distribution of X, consider any sequence of trial outcomes $s, s, \ldots, f$ – meaning that the first trial is a success, the second a success, ..., and the nth trial a failure – that results in i successes and $n - i$ failures. By independence, its probability of occurrence is $p \cdot p \cdots (1 - p) = p^i (1 - p)^{n-i}$. Because there are $\binom{n}{i} = \frac{n!}{(n-i)!i!}$ such sequences consisting of i values s and $n - i$ values f, it follows that

$$P(X = i) = \binom{n}{i} p^i (1 - p)^{n-i}, \quad i = 0, \ldots, n$$

Although we could compute the expected value of X by using the preceding to write

$$E[X] = \sum_{i=0}^{n} i P(X = i) = \sum_{i=0}^{n} i \binom{n}{i} p^i (1 - p)^{n-i}$$

and then attempt to simplify the preceding, it is easier to compute $E[X]$ by using the representation

$$X = \sum_{j=1}^{n} X_j,$$

where X_j is defined to equal 1 if trial j is a success and to equal 0 otherwise. Using Proposition 1.3.1, we obtain that

$$E[X] = \sum_{j=1}^{n} E[X_j] = np,$$

where the final equality used the result of Example 1.3c.   □

The following result will be used in Chapter 3.

Proposition 1.3.2 *Consider n independent trials, each of which is a success with probability p. Then, given that there is a total of i successes in the n trials, each of the $\binom{n}{i}$ subsets of i trials is equally likely to be the set of trials that resulted in successes.*

Proof. To verify the preceding, let T be any subset of size i of the set $\{1, \ldots, n\}$, and let A be the event that all of the trials in T were successes. Letting X be the number of successes in the n trials, then

$$P(A|X = i) = \frac{P(A, X = i)}{P(X = i)}$$

Now, $P(A, X = i)$ is the probability that all trials in T are successes and all trials not in T are failures. Consequently, on using the independence of the trials, we obtain from the preceding that

$$P(A|X = i) = \frac{p^i (1 - p)^{n-i}}{\binom{n}{i} p^i (1 - p)^{n-i}} = \frac{1}{\binom{n}{i}}$$

which proves the result.   □

The random variables $X_1, \ldots, X_n$ are said to be *independent* if probabilities concerning any subset of them are unchanged by information as to the values of the others.

Example 1.3e Suppose that k balls are to be randomly chosen from a set of N balls, of which n are red. If we let X_i equal 1 if the ith ball chosen is red and 0 if it is black, then $X_1, \ldots, X_n$ would be independent if each selected ball is replaced before the next selection is made, but they would not be independent if each selection is made without replacing previously selected balls. (Why not?)   □

Whereas the average of the possible values of X is indicated by its expected value, its spread is measured by its variance.

Definition The *variance* of X, denoted by $\text{Var}(X)$, is defined by

$$\text{Var}(X) = E[(X - E[X])^2].$$

In other words, the variance measures the average square of the difference between X and its expected value.

Example 1.3f Find $\text{Var}(X)$ when X is a Bernoulli random variable with parameter p.

Solution. Because $E[X] = p$ (as shown in Example 1.3c), we see that

$$(X - E[X])^2 = \begin{cases} (1 - p)^2 & \text{with probability } p \\ p^2 & \text{with probability } 1 - p. \end{cases}$$

Hence,

$$\text{Var}(X) = E[(X - E[X])^2]$$
$$= (1 - p)^2 p + p^2(1 - p)$$
$$= p - p^2. \qquad \square$$

If a and b are constants, then

$$\text{Var}(aX + b) = E[(aX + b - E[aX + b])^2]$$
$$= E[(aX - aE[X])^2] \quad \text{(by Equation (1.7))}$$
$$= E[a^2(X - E[X])^2]$$
$$= a^2 \text{Var}(X). \qquad (1.8)$$

Although it is not generally true that the variance of the sum of random variables is equal to the sum of their variances, this *is* the case when the random variables are independent.

Proposition 1.3.2 *If $X_1, \ldots, X_k$ are independent random variables, then*

$$\text{Var}\left(\sum_{j=1}^{k} X_j \right) = \sum_{j=1}^{k} \text{Var}(X_j).$$

Example 1.3g Find the variance of X, a binomial random variable with parameters n and p.

Solution. Recalling that X represents the number of successes in n *independent* trials (each of which is a success with probability p), we can represent it as

$$X = \sum_{j=1}^{n} X_j,$$

where X_j is defined to equal 1 if trial j is a success and 0 otherwise. Hence,

$$\text{Var}(X) = \sum_{j=1}^{n} \text{Var}(X_j) \quad \text{(by Proposition 1.3.2)}$$

$$= \sum_{j=1}^{n} p(1-p) \quad \text{(by Example 1.3f)}$$

$$= np(1-p). \qquad \square$$

The square root of the variance is called the *standard deviation*. As we shall see, a random variable tends to lie within a few standard deviations of its expected value.

1.4   Covariance and Correlation

The covariance of any two random variables X and Y, denoted by $\text{Cov}(X, Y)$, is defined by

$$\text{Cov}(X, Y) = E[(X - E[X])(Y - E[Y])].$$

Upon multiplying the terms within the expectation, and then taking expectation term by term, it can be shown that

$$\text{Cov}(X, Y) = E[XY] - E[X]E[Y].$$

A positive value of the covariance indicates that X and Y both tend to be large at the same time, whereas a negative value indicates that when one is large the other tends to be small. (Independent random variables have covariance equal to 0.)

Example 1.4a Let X and Y both be Bernoulli random variables. That is, each takes on either the value 0 or 1. Using the identity

$$\text{Cov}(X, Y) = E[XY] - E[X]E[Y]$$

and noting that XY will equal 1 or 0 depending upon whether both X and Y are equal to 1, we obtain that

$$\text{Cov}(X, Y) = P\{X = 1, \ Y = 1\} - P\{X = 1\}P\{Y = 1\}.$$

From this, we see that

$$\text{Cov}(X, Y) > 0 \iff P\{X = 1, Y = 1\} > P\{X = 1\}P\{Y = 1\}$$

$$\iff \frac{P\{X = 1, Y = 1\}}{P\{X = 1\}} > P\{Y = 1\}$$

$$\iff P\{Y = 1 \mid X = 1\} > P\{Y = 1\}.$$

That is, the covariance of X and Y is positive if the outcome that $X = 1$ makes it more likely that $Y = 1$ (which, as is easily seen, also implies the reverse). $\qquad\square$

The following properties of covariance are easily established. For random variables X and Y, and constant c:

$$\text{Cov}(X, Y) = \text{Cov}(Y, X),$$

$$\text{Cov}(X, X) = \text{Var}(X),$$

$$\text{Cov}(cX, Y) = c\,\text{Cov}(X, Y),$$

$$\text{Cov}(c, Y) = 0.$$

Covariance, like expected value, satisfies a linearity property – namely,

$$\text{Cov}(X_1 + X_2, Y) = \text{Cov}(X_1, Y) + \text{Cov}(X_2, Y). \qquad (1.9)$$

Equation (1.9) is proven as follows:

$$\text{Cov}(X_1 + X_2, Y) = E[(X_1 + X_2)Y] - E[X_1 + X_2]E[Y]$$

$$= E[X_1Y + X_2Y] - (E[X_1] + E[X_2])E[Y]$$

$$= E[X_1Y] - E[X_1]E[Y] + E[X_2Y] - E[X_2]E[Y]$$

$$= \text{Cov}(X_1, Y) + \text{Cov}(X_2, Y).$$

Equation (1.9) is easily generalized to yield the following useful identity:

$$\text{Cov}\left(\sum_{i=1}^{n} X_i, \sum_{j=1}^{m} Y_j\right) = \sum_{i=1}^{n}\sum_{j=1}^{m}\text{Cov}(X_i, Y_j). \qquad (1.10)$$

Equation (1.10) yields a useful formula for the variance of the sum of random variables:

$$\text{Var}\left(\sum_{i=1}^{n} X_i\right) = \text{Cov}\left(\sum_{i=1}^{n} X_i, \sum_{j=1}^{n} X_j\right)$$

$$= \sum_{i=1}^{n} \sum_{j=1}^{n} \text{Cov}(X_i, X_j)$$

$$= \sum_{i=1}^{n} \text{Cov}(X_i, X_i) + \sum_{i=1}^{n} \sum_{j\neq i} \text{Cov}(X_i, X_j)$$

$$= \sum_{i=1}^{n} \text{Var}(X_i) + \sum_{i=1}^{n} \sum_{j\neq i} \text{Cov}(X_i, X_j). \tag{1.11}$$

The degree to which large values of X tend to be associated with large values of Y is measured by the *correlation* between X and Y, denoted as $\rho(X, Y)$ and defined by

$$\rho(X, Y) = \frac{\text{Cov}(X, Y)}{\sqrt{\text{Var}(X)\,\text{Var}(Y)}}.$$

It can be shown that

$$-1 \leq \rho(X, Y) \leq 1.$$

If X and Y are linearly related by the equation

$$Y = a + bX,$$

then $\rho(X, Y)$ will equal 1 when b is positive and -1 when b is negative.

1.5   Conditional Expectation

For random variables X and Y, we define the conditional expectation of X given that $Y = y$ by

$$E[X|Y = y] = \sum_{x} x P(X = x|Y = y)$$

That is, the conditional expectation of X given that $Y = y$ is, like the ordinary expectation of X, a weighted average of the possible values of X; but now the value x is weighted not by the unconditional probability that $X = x$, but by its conditional probability given the information that $Y = y$.

An important property of conditional expectation is that the expected value of X is a weighted average of the conditional expectation of X given that $Y = y$. That is, we have the following:

Proposition 1.5.1

$$E[X] = \sum_y E[X|Y = y]P(Y = y)$$

Proof.

$$\sum_y E[X|Y = y]P(Y = y) = \sum_y \sum_x x P(X = x|Y = y)P(Y = y)$$

$$= \sum_y \sum_x x P(X = x, Y = y)$$

$$= \sum_x x \sum_y P(X = x, Y = y)$$

$$= \sum_x x P(X = x)$$

$$= E[X] \qquad \qquad \square$$

Let $E[X|Y]$ be that function of the random variable Y which, when $Y = y$, is defined to equal $E[X|Y = y]$. Using that the expected value of any function of Y, say $h(Y)$, can be expressed as (see Exercise 1.20)

$$E[h(Y)] = \sum_y h(y)P(Y = y)$$

it follows that

$$E[E[X|Y]] = \sum_y E[X|Y = y]P(Y = y)$$

Hence, the preceding proposition can be written as

$$E[X] = E[E[X|Y]]$$

1.6   Exercises

Exercise 1.1 When typing a report, a certain typist makes i errors with probability p_i $(i \geq 0)$, where

$$p_0 = .20, \quad p_1 = .35, \quad p_2 = .25, \quad p_3 = .15.$$

What is the probability that the typist makes

(a) at least four errors;
(b) at most two errors?

Exercise 1.2 A family picnic scheduled for tomorrow will be postponed if it is either cloudy or rainy. If the probability that it will be cloudy is .40, the probability that it will be rainy is .30, and the probability that it will be both rainy and cloudy is .20, what is the probabilty that the picnic will not be postponed?

Exercise 1.3 If two people are randomly chosen from a group of eight women and six men, what is the probability that

(a) both are women;
(b) both are men;
(c) one is a man and the other a woman?

Exercise 1.4 A club has 120 members, of whom 35 play chess, 58 play bridge, and 27 play both chess and bridge. If a member of the club is randomly chosen, what is the conditional probability that she

(a) plays chess given that she plays bridge;
(b) plays bridge given that she plays chess?

Exercise 1.5 Cystic fibrosis (CF) is a genetically caused disease. A child that receives a CF gene from each of its parents will develop the disease either as a teenager or before, and will not live to adulthood. A child that receives either zero or one CF gene will not develop the disease. If an individual has a CF gene, then each of his or her children will independently receive that gene with probability $1/2$.

(a) If both parents possess the CF gene, what is the probability that their child will develop cystic fibrosis?
(b) What is the probability that a 30-year old who does not have cystic fibrosis, but whose sibling died of that disease, possesses a CF gene?

Exercise 1.6 Two cards are randomly selected from a deck of 52 playing cards. What is the conditional probability they are both aces, given that they are of different suits?

Exercise 1.7 If A and B are independent, show that so are

(a) A and B^c;
(b) A^c and B^c.

Exercise 1.8 A gambling book recommends the following strategy for the game of roulette. It recommends that the gambler bet 1 on red. If red appears (which has probability 18/38 of occurring) then the gambler should take his profit of 1 and quit. If the gambler loses this bet, he should then make a second bet of size 2 and then quit. Let X denote the gambler's winnings.

(a) Find $P\{X > 0\}$.
(b) Find $E[X]$.

Exercise 1.9 Four buses carrying 152 students from the same school arrive at a football stadium. The buses carry (respectively) 39, 33, 46, and 34 students. One of the 152 students is randomly chosen. Let X denote the number of students who were on the bus of the selected student. One of the four bus drivers is also randomly chosen. Let Y be the number of students who were on that driver's bus.

(a) Which do you think is larger, $E[X]$ or $E[Y]$?
(b) Find $E[X]$ and $E[Y]$.

Exercise 1.10 Two players play a tennis match, which ends when one of the players has won two sets. Suppose that each set is equally likely to be won by either player, and that the results from different sets are independent. Find (a) the expected value and (b) the variance of the number of sets played.

Exercise 1.11 Verify that

$$\text{Var}(X) = E[X^2] - (E[X])^2.$$

Hint: Starting with the definition

$$\text{Var}(X) = E[(X - E[X])^2],$$

square the expression on the right side; then use the fact that the expected value of a sum of random variables is equal to the sum of their expectations.

Exercise 1.12 A lawyer must decide whether to charge a fixed fee of $5,000 or take a contingency fee of $25,000 if she wins the case (and 0 if she loses). She estimates that her probability of winning is .30. Determine the mean and standard deviation of her fee if

(a) she takes the fixed fee;
(b) she takes the contingency fee.

Exercise 1.13 Let $X_1, \ldots, X_n$ be independent random variables, all having the same distribution with expected value μ and variance σ^2. The random variable $\bar{X}$, defined as the arithmetic average of these variables, is called the *sample mean*. That is, the sample mean is given by

$$\bar{X} = \frac{\sum_{i=1}^{n} X_i}{n}.$$

(a) Show that $E[\bar{X}] = \mu$.
(b) Show that $\mathrm{Var}(\bar{X}) = \sigma^2/n$.

The random variable S^2, defined by

$$S^2 = \frac{\sum_{i=1}^{n}(X_i - \bar{X})^2}{n-1},$$

is called the *sample variance.*

(c) Show that $\sum_{i=1}^{n}(X_i - \bar{X})^2 = \sum_{i=1}^{n} X_i^2 - n\bar{X}^2$.
(d) Show that $E[S^2] = \sigma^2$.

Exercise 1.14 Verify that

$$\mathrm{Cov}(X, Y) = E[XY] - E[X]E[Y].$$

Exercise 1.15 Prove:

(a) $\mathrm{Cov}(X, Y) = \mathrm{Cov}(Y, X)$;
(b) $\mathrm{Cov}(X, X) = \mathrm{Var}(X)$;
(c) $\mathrm{Cov}(cX, Y) = c\,\mathrm{Cov}(X, Y)$;
(d) $\mathrm{Cov}(c, Y) = 0$.

Exercise 1.16 If U and V are independent random variables, both having variance 1, find $\mathrm{Cov}(X, Y)$ when

$$X = aU + bV, \qquad Y = cU + dV.$$

Exercise 1.17 If $\text{Cov}(X_i, X_j) = ij$, find

(a) $\text{Cov}(X_1 + X_2, X_3 + X_4)$;
(b) $\text{Cov}(X_1 + X_2 + X_3, X_2 + X_3 + X_4)$.

Exercise 1.18 Suppose that – in any given time period – a certain stock is equally likely to go up 1 unit or down 1 unit, and that the outcomes of different periods are independent. Let X be the amount the stock goes up (either 1 or -1) in the first period, and let Y be the cumulative amount it goes up in the first three periods. Find the correlation between X and Y.

Exercise 1.19 Can you construct a pair of random variables such that $\text{Var}(X) = \text{Var}(Y) = 1$ and $\text{Cov}(X, Y) = 2$?

Exercise 1.20 If Y is a random variable and h a function, then $h(Y)$ is also a random variable. If the set of distinct possible values of $h(Y)$ are $\{h_i, i \geq 1\}$, then by the definition of expected value, we have that $E[h(Y)] = \sum_i h_i P(h(Y) = h_i)$. On the other hand, because $h(Y)$ is equal to $h(y)$ when $Y = y$, it is intuitive that

$$E[h(Y)] = \sum_y h(y) P(Y = y)$$

Verify that the preceding equation is valid.

Exercise 1.21 The *distribution function* $F(x)$ of the random variable X is defined by

$$F(x) = P(X \leq x)$$

If X takes on one of the values $1, 2, \ldots$, and F is a known function, how would you obtain $P(X = i)$?

REFERENCE

[1] Ross, S. M. (2010). *A First Course in Probability*, 8th ed. Englewood Cliffs, NJ: Prentice-Hall.

2. Normal Random Variables

2.1   Continuous Random Variables

Whereas the possible values of the random variables considered in the previous chapter constituted sets of discrete values, there exist random variables whose set of possible values is instead a continuous region. These *continuous* random variables can take on any value within some interval. For example, such random variables as the time it takes to complete an assignment, or the weight of a randomly chosen individual, are usually considered to be continuous.

Every continuous random variable X has a function f associated with it. This function, called the *probability density function* of X, determines the probabilities associated with X in the following manner. For any numbers $a < b$, the area under f between a and b is equal to the probability that X assumes a value between a and b. That is,

$$P\{a \le X \le b\} = \text{area under } f \text{ between } a \text{ and } b.$$

Figure 2.1 presents a probability density function.

2.2   Normal Random Variables

A very important type of continuous random variable is the normal random variable. The probability density function of a normal random variable X is determined by two parameters, denoted by μ and σ, and is given by the formula

$$f(x) = \frac{1}{\sqrt{2\pi}\sigma} e^{-(x-\mu)^2/2\sigma^2}, \quad -\infty < x < \infty.$$

A plot of the normal probability density function gives a bell-shaped curve that is symmetric about the value μ, and with a variability that is measured by σ. The larger the value of σ, the more spread there is in f. Figure 2.2 presents three different normal probability density functions. Note how the curve flattens out as σ increases.

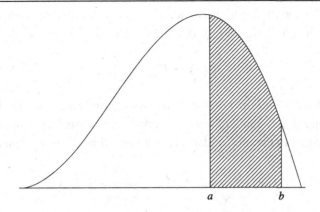

$P\{a \leq X \leq b\}$ = area of shaded region

Figure 2.1: Probability Density Function of X

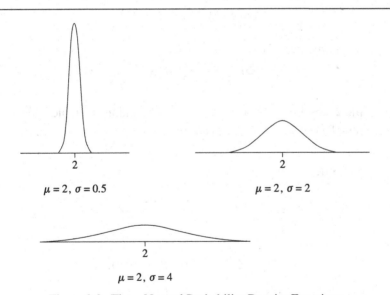

$\mu = 2, \sigma = 0.5$ $\mu = 2, \sigma = 2$

$\mu = 2, \sigma = 4$

Figure 2.2: Three Normal Probability Density Functions

It can be shown that the parameters μ and σ^2 are equal to the expected value and to the variance of X, respectively. That is,

$$\mu = E[X], \qquad \sigma^2 = \text{Var}(X).$$

A normal random variable having mean 0 and variance 1 is called a *standard normal* random variable. Let Z be a standard normal random variable. The function $\Phi(x)$, defined for all real numbers x by

$$\Phi(x) = P\{Z \leq x\},$$

is called the *standard normal distribution function*. Thus $\Phi(x)$, the probability that a standard normal random variable is less than or equal to x, is equal to the area under the *standard normal density function*

$$f(x) = \frac{1}{\sqrt{2\pi}}e^{-x^2/2}, \qquad -\infty < x < \infty,$$

between $-\infty$ and x. Table 2.1 specifies values of $\Phi(x)$ when $x > 0$. Probabilities for negative x can be obtained by using the symmetry of the standard normal density about 0 to conclude (see Figure 2.3) that

$$P\{Z < -x\} = P\{Z > x\}$$

or, equivalently, that

$$\Phi(-x) = 1 - \Phi(x).$$

Example 2.2a Let Z be a standard normal random variable. For $a < b$, express $P\{a < Z \leq b\}$ in terms of Φ.

Solution. Since

$$P\{Z \leq b\} = P\{Z \leq a\} + P\{a < Z \leq b\},$$

we see that

$$P\{a < Z \leq b\} = \Phi(b) - \Phi(a). \qquad \square$$

Example 2.2b Tabulated values of $\Phi(x)$ show that, to four decimal places,

$$P\{|Z| \leq 1\} = P\{-1 \leq Z \leq 1\} = .6826,$$
$$P\{|Z| \leq 2\} = P\{-2 \leq Z \leq 2\} = .9544,$$
$$P\{|Z| \leq 3\} = P\{-3 \leq Z \leq 3\} = .9974. \qquad \square$$

Table 2.1: $\Phi(x) = P\{Z \le x\}$

x	.00	.01	.02	.03	.04	.05	.06	.07	.08	.09
0.0	.5000	.5040	.5080	.5120	.5160	.5199	.5239	.5279	.5319	.5359
0.1	.5398	.5438	.5478	.5517	.5557	.5596	.5636	.5675	.5714	.5753
0.2	.5793	.5832	.5871	.5910	.5948	.5987	.6026	.6064	.6103	.6141
0.3	.6179	.6217	.6255	.6293	.6331	.6368	.6406	.6443	.6480	.6517
0.4	.6554	.6591	.6628	.6664	.6700	.6736	.6772	.6808	.6844	.6879
0.5	.6915	.6950	.6985	.7019	.7054	.7088	.7123	.7157	.7190	.7224
0.6	.7257	.7291	.7324	.7357	.7389	.7422	.7454	.7486	.7517	.7549
0.7	.7580	.7611	.7642	.7673	.7704	.7734	.7764	.7794	.7823	.7852
0.8	.7881	.7910	.7939	.7967	.7995	.8023	.8051	.8078	.8106	.8133
0.9	.8159	.8186	.8212	.8238	.8264	.8289	.8315	.8340	.8365	.8389
1.0	.8413	.8438	.8461	.8485	.8508	.8531	.8554	.8577	.8599	.8621
1.1	.8643	.8665	.8686	.8708	.8729	.8749	.8770	.8790	.8810	.8830
1.2	.8849	.8869	.8888	.8907	.8925	.8944	.8962	.8980	.8997	.9015
1.3	.9032	.9049	.9066	.9082	.9099	.9115	.9131	.9147	.9162	.9177
1.4	.9192	.9207	.9222	.9236	.9251	.9265	.9279	.9292	.9306	.9319
1.5	.9332	.9345	.9357	.9370	.9382	.9394	.9406	.9418	.9429	.9441
1.6	.9452	.9463	.9474	.9484	.9495	.9505	.9515	.9525	.9535	.9545
1.7	.9554	.9564	.9573	.9582	.9591	.9599	.9608	.9616	.9625	.9633
1.8	.9641	.9649	.9656	.9664	.9671	.9678	.9686	.9693	.9699	.9706
1.9	.9713	.9719	.9726	.9732	.9738	.9744	.9750	.9756	.9761	.9767
2.0	.9772	.9778	.9783	.9788	.9793	.9798	.9803	.9808	.9812	.9817
2.1	.9821	.9826	.9830	.9834	.9838	.9842	.9846	.9850	.9854	.9857
2.2	.9861	.9864	.9868	.9871	.9875	.9878	.9881	.9884	.9887	.9890
2.3	.9893	.9896	.9898	.9901	.9904	.9906	.9909	.9911	.9913	.9916
2.4	.9918	.9920	.9922	.9925	.9927	.9929	.9931	.9932	.9934	.9936
2.5	.9938	.9940	.9941	.9943	.9945	.9946	.9948	.9949	.9951	.9952
2.6	.9953	.9955	.9956	.9957	.9959	.9960	.9961	.9962	.9963	.9964
2.7	.9965	.9966	.9967	.9968	.9969	.9970	.9971	.9972	.9973	.9974
2.8	.9974	.9975	.9976	.9977	.9977	.9978	.9979	.9979	.9980	.9981
2.9	.9981	.9982	.9982	.9983	.9984	.9984	.9985	.9985	.9986	.9986
3.0	.9987	.9987	.9987	.9988	.9988	.9989	.9989	.9989	.9990	.9990
3.1	.9990	.9991	.9991	.9991	.9992	.9992	.9992	.9992	.9993	.9993
3.2	.9993	.9993	.9994	.9994	.9994	.9994	.9994	.9995	.9995	.9995
3.3	.9995	.9995	.9995	.9996	.9996	.9996	.9996	.9996	.9996	.9997
3.4	.9997	.9997	.9997	.9997	.9997	.9997	.9997	.9997	.9997	.9998

When greater accuracy than that provided by Table 2.1 is needed, the following approximation to $\Phi(x)$, accurate to six decimal places, can be used: For $x > 0$,

$$\Phi(x) \approx 1 - \frac{1}{\sqrt{2\pi}}e^{-x^2/2}(a_1 y + a_2 y^2 + a_3 y^3 + a_4 y^4 + a_5 y^5),$$

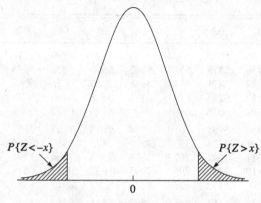

Figure 2.3: $P\{Z < -x\} = P\{Z > x\}$

where

$$y = \frac{1}{1 + .2316419x},$$

$$a_1 = .319381530,$$

$$a_2 = -.356563782,$$

$$a_3 = 1.781477937,$$

$$a_4 = -1.821255978,$$

$$a_5 = 1.330274429,$$

and

$$\Phi(-x) = 1 - \Phi(x).$$

2.3   Properties of Normal Random Variables

An important property of normal random variables is that if X is a normal random variable then so is $aX + b$, when a and b are constants. This property enables us to transform any normal random variable X into a standard normal random variable. For suppose X is normal with mean μ and variance σ^2. Then, since (from Equations (1.7) and (1.8))

$$Z = \frac{X - \mu}{\sigma}$$

has expected value 0 and variance 1, it follows that Z is a standard normal random variable. As a result, we can compute probabilities for any normal random variable in terms of the standard normal distribution function Φ.

Example 2.3a IQ examination scores for sixth-graders are normally distributed with mean value 100 and standard deviation 14.2. What is the probability that a randomly chosen sixth-grader has an IQ score greater than 130?

Solution. Let X be the score of a randomly chosen sixth-grader. Then,

$$P\{X > 130\} = P\left\{\frac{X - 100}{14.2} > \frac{130 - 100}{14.2}\right\}$$

$$= P\left\{\frac{X - 100}{14.2} > 2.113\right\}$$

$$= 1 - \Phi(2.113)$$

$$= .017. \qquad \square$$

Example 2.3b Let X be a normal random variable with mean μ and standard deviation σ. Then, since

$$|X - \mu| \le a\sigma$$

is equivalent to

$$\left|\frac{X - \mu}{\sigma}\right| \le a,$$

it follows from Example 2.2b that 68.26% of the time a normal random variable will be within one standard deviation of its mean; 95.44% of the time it will be within two standard deviations of its mean; and 99.74% of the time it will be within three standard deviations of its mean. $\square$

Another important property of normal random variables is that the sum of independent normal random variables is also a normal random variable. That is, if X_1 and X_2 are independent normal random variables with means μ_1 and μ_2 and with standard deviations σ_1 and σ_2, then $X_1 + X_2$ is normal with mean

$$E[X_1 + X_2] = E[X_1] + E[X_2] = \mu_1 + \mu_2$$

and variance

$$\text{Var}(X_1 + X_2) = \text{Var}(X_1) + \text{Var}(X_2) = \sigma_1^2 + \sigma_2^2.$$

Example 2.3c The annual rainfall in Cleveland, Ohio, is normally distributed with mean 40.14 inches and standard deviation 8.7 inches. Find the probabiity that the sum of the next two years' rainfall exceeds 84 inches.

Solution. Let X_i denote the rainfall in year i ($i = 1, 2$). Then, assuming that the rainfalls in successive years can be assumed to be independent, it follows that $X_1 + X_2$ is normal with mean 80.28 and variance $2(8.7)^2 = 151.38$. Therefore, with Z denoting a standard normal random variable,

$$P\{X_1 + X_2 > 84\} = P\left\{Z > \frac{84 - 80.28}{\sqrt{151.38}}\right\}$$

$$= P\{Z > .3023\}$$

$$\approx .3812. \qquad \square$$

The random variable Y is said to be a *lognormal* random variable with parameters μ and σ if $\log(Y)$ is a normal random variable with mean μ and variance σ^2. That is, Y is lognormal if it can be expressed as

$$Y = e^X,$$

where X is a normal random variable. The mean and variance of a lognormal random variable are as follows:

$$E[Y] = e^{\mu + \sigma^2/2},$$

$$\text{Var}(Y) = e^{2\mu + 2\sigma^2} - e^{2\mu + \sigma^2} = e^{2\mu + \sigma^2}(e^{\sigma^2} - 1).$$

Example 2.3d Starting at some fixed time, let $S(n)$ denote the price of a certain security at the end of n additional weeks, $n \geq 1$. A popular model for the evolution of these prices assumes that the price ratios $S(n)/S(n-1)$ for $n \geq 1$ are independent and identically distributed (i.i.d.) lognormal random variables. Assuming this model, with lognormal parameters $\mu = .0165$ and $\sigma = .0730$, what is the probability that

(a) the price of the security increases over each of the next two weeks;
(b) the price at the end of two weeks is higher than it is today?

Solution. Let Z be a standard normal random variable. To solve part (a), we use that $\log(x)$ increases in x to conclude that $x > 1$ if and only if $\log(x) > \log(1) = 0$. As a result, we have

$$P\left\{\frac{S(1)}{S(0)} > 1\right\} = P\left\{\log\left(\frac{S(1)}{S(0)}\right) > 0\right\}$$

$$= P\left\{Z > \frac{-.0165}{.0730}\right\}$$

$$= P\{Z > -.2260\}$$

$$= P\{Z < .2260\}$$

$$\approx .5894.$$

Therefore, the probability that the price is up after one week is .5894. Since the successive price ratios are independent, the probability that the price increases over each of the next two weeks is $(.5894)^2 = .3474$.

To solve part (b), reason as follows:

$$P\left\{\frac{S(2)}{S(0)} > 1\right\} = P\left\{\frac{S(2)}{S(1)}\frac{S(1)}{S(0)} > 1\right\}$$

$$= P\left\{\log\left(\frac{S(2)}{S(1)}\right) + \log\left(\frac{S(1)}{S(0)}\right) > 0\right\}$$

$$= P\left\{Z > \frac{-.0330}{.0730\sqrt{2}}\right\}$$

$$= P\{Z > -.31965\}$$

$$= P\{Z < .31965\}$$

$$\approx .6254,$$

where we have used that $\log\left(\frac{S(2)}{S(1)}\right) + \log\left(\frac{S(1)}{S(0)}\right)$, being the sum of independent normal random variables with a common mean .0165 and a common standard deviation .0730, is itself a normal random variable with mean .0330 and variance $2(.0730)^2$. $\qquad\square$

2.4   The Central Limit Theorem

The ubiquity of normal random variables is explained by the central limit theorem, probably the most important theoretical result in probability.

This theorem states that the sum of a large number of independent random variables, all having the same probability distribution, will itself be approximately a normal random variable.

For a more precise statement of the central limit theorem, suppose that $X_1, X_2, \ldots$ is a sequence of i.i.d. random variables, each with expected value μ and variance σ^2, and let

$$S_n = \sum_{i=1}^{n} X_i.$$

Central Limit Theorem *For large n, S_n will approximately be a normal random variable with expected value $n\mu$ and variance $n\sigma^2$. As a result, for any x we have*

$$P\left\{ \frac{S_n - n\mu}{\sigma\sqrt{n}} \leq x \right\} \approx \Phi(x),$$

with the approximation becoming exact as n becomes larger and larger.

Suppose that X is a binomial random variable with parameters n and p. Since X represents the number of successes in n independent trials, each of which is a success with probability p, it can be expressed as

$$X = \sum_{i=1}^{n} X_i,$$

where X_i is 1 if trial i is a success and is 0 otherwise. Since (from Section 1.3)

$$E[X_i] = p \quad \text{and} \quad \text{Var}(X_i) = p(1 - p),$$

it follows from the central limit theorem that, when n is large, X will approximately have a normal distribution with mean np and variance $np(1 - p)$.

Example 2.4a A fair coin is tossed 100 times. What is the probability that heads appears fewer than 40 times?

Solution. If X denotes the number of heads, then X is a binomial random variable with parameters $n = 100$ and $p = 1/2$. Since $np = 50$ we have $np(1 - p) = 25$, and so

$$P\{X < 40\} = P\left\{\frac{X - 50}{\sqrt{25}} < \frac{40 - 50}{\sqrt{25}}\right\}$$

$$= P\left\{\frac{X - 50}{\sqrt{25}} < -2\right\}$$

$$\approx \Phi(-2)$$

$$= .0228.$$

A computer program for computing binomial probabilities gives the exact solution .0176, and so the preceding is not quite as acccurate as we might like. However, we could improve the approximation by noting that, since X is an integral-valued random variable, the event that $X < 40$ is equivalent to the event that $X < 39 + c$ for any c, $0 < c \leq 1$. Consequently, a better approximation may be obtained by writing the desired probability as $P\{X < 39.5\}$. This gives

$$P\{X < 39.5\} = P\left\{\frac{X - 50}{\sqrt{25}} < \frac{39.5 - 50}{\sqrt{25}}\right\}$$

$$= P\left\{\frac{X - 50}{\sqrt{25}} < -2.1\right\}$$

$$\approx \Phi(-2.1)$$

$$= .0179,$$

which is indeed a better approximation.   □

2.5   Exercises

Exercise 2.1 For a standard normal random variable Z, find:

(a) $P\{Z < -.66\}$;
(b) $P\{|Z| < 1.64\}$;
(c) $P\{|Z| > 2.20\}$.

Exercise 2.2 Find the value of x when Z is a standard normal random variable and

$$P\{-2 < Z < -1\} = P\{1 < Z < x\}.$$

Exercise 2.3 Argue (a picture is acceptable) that

$$P\{|Z| > x\} = 2P\{Z > x\},$$

where $x > 0$ and Z is a standard normal random variable.

Exercise 2.4 Let X be a normal random variable having expected value μ and variance σ^2, and let $Y = a + bX$. Find values a, b ($a \neq 0$) that give Y the same distribution as X. Then, using these values, find $Cov(X, Y)$.

Exercise 2.5 The systolic blood pressure of male adults is normally distributed with a mean of 127.7 and a standard deviation of 19.2.

(a) Specify an interval in which the blood pressures of approximately 68% of the adult male population fall.
(b) Specify an interval in which the blood pressures of approximately 95% of the adult male population fall.
(c) Specify an interval in which the blood pressures of approximately 99.7% of the adult male population fall.

Exercise 2.6 Suppose that the amount of time that a certain battery functions is a normal random variable with mean 400 hours and standard deviation 50 hours. Suppose that an individual owns two such batteries, one of which is to be used as a spare to replace the other when it fails.

(a) What is the probability that the total life of the batteries will exceed 760 hours?
(b) What is the probability that the second battery will outlive the first by at least 25 hours?
(c) What is the probability that the longer-lasting battery will outlive the other by at least 25 hours?

Exercise 2.7 The time it takes to develop a photographic print is a random variable with mean 18 seconds and standard deviation 1 second. Approximate the probability that the total amount of time that it takes to process 100 prints is

(a) more than 1,710 seconds;
(b) between 1,690 and 1,710 seconds.

Exercise 2.8 Frequent fliers of a certain airline fly a random number of miles each year, having mean and standard deviation of 25,000 and 12,000 miles, respectively. If 30 such people are randomly chosen, approximate the probability that the average of their mileages for this year will

(a) exceed 25,000;
(b) be between 23,000 and 27,000.

Exercise 2.9 A model for the movement of a stock supposes that, if the present price of the stock is s, then – after one time period – it will either be us with probability p or ds with probability $1 - p$. Assuming that successive movements are independent, approximate the probability that the stock's price will be up at least 30% after the next 1,000 time periods if $u = 1.012$, $d = .990$, and $p = .52$.

Exercise 2.10 In each time period, a certain stock either goes down 1 with probability .39, remains the same with probability .20, or goes up 1 with probability .41. Asuming that the changes in successive time periods are independent, approximate the probability that, after 700 time periods, the stock will be up more than 10 from where it started.

REFERENCE

[1] Ross, S. M. (2010). *A First Course in Probability,* Prentice-Hall.

3. Brownian Motion and Geometric Brownian Motion

3.1 Brownian Motion

A Brownian motion is a collection of random variables $X(t), t \geq 0$ that satisfy certain properties that we will momentarily present. We imagine that we are observing some process as it evolves over time. The index parameter t represents time, and $X(t)$ is interpreted as the state of the process at time t. Here is a formal definition.

Definition The collection of random variables $X(t), t \geq 0$ is said to be a *Brownian motion* with drift parameter μ and variance parameter σ^2 if the following hold:

(a) $X(0)$ is a given constant.
(b) For all positive y and t, the random variable $X(t + y) - X(y)$ is independent of the the process values up to time y and has a normal distribution with mean μt and variance $t\sigma^2$.

Assumption (b) says that, for any history of the process up to the present time y, the change in the value of the process over the next t time units is a normal random with mean μt and variance $t\sigma^2$. Because any future value $X(t + y)$ is equal to the present value $X(y)$ plus the change in value $X(t + y) - X(y)$, the assumption implies that it is only the present value of the process, and not any past values, that determines probabilities about future values.

An important property of Brownian motion is that $X(t)$ will, with probability 1, be a continuous function of t. Althought this is a mathematically deep result, it is not difficult to see why it might be true. To prove that $X(t)$ is continuous, we must show that

$$\lim_{h \to 0} (X(t + h) - X(t)) = 0$$

However, because the random variable $X(t + h) - X(t)$ has mean μh and variance $h\sigma^2$, it converges as $h \to 0$ to a random variable with mean

0 and variance 0. That is, it converges to the constant 0, thus arguing for continuity.

Although $X(t)$ will, with probability 1, be a continuous function of t, it possesses the startling property of being nowhere differentiable. To see why this might be the case, note that $\frac{X(t+h)-X(t)}{h}$ has mean μ and variance σ^2/h. Because the variance of this ratio is converging to infinity as $h \to 0$, it is not surprising that the ratio does not converge.

3.2   Brownian Motion as a Limit of Simpler Models

Let Δ be a small increment of time, and consider a process such that every Δ time units the value of the process either increases by the amount $\sigma\sqrt{\Delta}$ with probability p or decreases by the amount $\sigma\sqrt{\Delta}$ with probability $1 - p$, where

$$p = \frac{1}{2}\left(1 + \frac{\mu}{\sigma}\sqrt{\Delta}\right)$$

and where the successive changes in value are independent.

Thus, we are supposing that the process values change only at times that are integral multiples of Δ, and that at each change point the value of the process either increases or decreases by the amount $\sigma\sqrt{\Delta}$, with the change being an increase with probability $p = \frac{1}{2}(1 + \frac{\mu}{\sigma}\sqrt{\Delta})$.

As we take Δ smaller and smaller, so that changes occur more and more frequently (though by amounts that become smaller and smaller), the process becomes a Brownian motion with drift parameter μ and variance parameter σ^2. Consequently, Brownian motion can be approximated by a relatively simple process that either increases or decreases by a fixed amount at regularly specified times.

We now verify that the preceding model becomes Brownian motion as we let Δ become smaller and smaller. To begin, let

$$X_i = \begin{cases} 1, & \text{if the change at time } i\Delta \text{ is an increase} \\ -1, & \text{if the change at time } i\Delta \text{ is a decrease} \end{cases}$$

Hence, if $X(0)$ is the process value at time 0, then its value after n changes is

$$X(n\Delta) = X(0) + \sigma\sqrt{\Delta}\,(X_1 + \ldots + X_n)$$

Because there would have been $n = t/\Delta$ changes by time t, this gives that

$$X(t) - X(0) = \sigma\sqrt{\Delta}\sum_{i=1}^{t/\Delta} X_i$$

Because the X_i, $i = 1, \ldots, t/\Delta$, are independent, and as Δ goes to 0 there are more and more terms in the summation $\sum_{i=1}^{t/\Delta} X_i$, the central limit theorem suggests that this sum converges to a normal random variable. Consequently, as Δ goes to 0, the process value at time t becomes a normal random variable. To compute its mean and variance, note first that

$$E[X_i] = 1(p) - 1(1-p) = 2p - 1 = \frac{\mu}{\sigma}\sqrt{\Delta}$$

and

$$\text{Var}(X_i) = E\left[X_i^2\right] - (E[X_i])^2 = 1 - (2p-1)^2$$

Hence,

$$E[X(t) - X(0)] = E\left[\sigma\sqrt{\Delta}\sum_{i=1}^{t/\Delta} X_i\right]$$

$$= \sigma\sqrt{\Delta}\sum_{i=1}^{t/\Delta} E[X_i]$$

$$= \sigma\sqrt{\Delta}\,\frac{t}{\Delta}\,\frac{\mu}{\sigma}\sqrt{\Delta}$$

$$= \mu t$$

Furthermore,

$$\text{Var}(X(t) - X(0)) = \text{Var}\left(\sigma\sqrt{\Delta}\sum_{i=1}^{t/\Delta} X_i\right)$$

$$= \sigma^2\Delta\sum_{i=1}^{t/\Delta}\text{Var}(X_i) \quad \text{(by independence)}$$

$$= \sigma^2 t\left[1 - (2p-1)^2\right]$$

Because $p \to 1/2$ as $\Delta \to 0$, the preceding shows that

$$\text{Var}(X(t) - X(0)) \to t\sigma^2 \quad \text{as} \quad \Delta \to 0$$

Consequently, as Δ gets smaller and smaller, $X(t) - X(0)$ converges to a normal random variable with mean μt and variance $t\sigma^2$. In addition, because successive process changes are independent and each has the same probability of being an increase, it follows that $X(t + y) - X(y)$ has the same distribution as does $X(t) - X(0)$ and is, in addition, independent of earlier process changes before time y. Hence, it follows that as Δ goes to 0, the collection of process values over time becomes a Brownian motion process with drift parameter μ and variance parameter σ^2.

An important result about Brownian motion is that, conditional on the value of the process at time t, the joint distribution of the process values up to time t does not depend on the value of the drift parameter. This result is easily proven by using the approximating processes, as we now show.

Theorem 3.2.1 *Given that $X(t) = x$, the conditional probability law of the collection of prices $X(y), 0 \leq y \leq t$, is the same for all values of μ.*

Proof. Let $s = X(0)$ be the price at time 0. Now, consider the approximating model where the price changes every Δ time units by an amount equal, in absolute value, to $c \equiv \sigma\sqrt{\Delta}$, and note that c does not depend on μ. By time t, there would have been t/Δ changes. Hence, given that the price has increased from time 0 to time t by the amount $x - s$, it follows that, of the t/Δ changes, there have been a total of $\frac{t}{2\Delta} + \frac{x-s}{2c}$ positive changes and a total of $\frac{t}{2\Delta} - \frac{x-s}{2c}$ negative changes. (This follows because if the preceding were so, then, of the first t/Δ changes, there would have been $\frac{x-s}{c}$ more positive than negative changes, and so the price would have increased by $c(\frac{x-s}{c}) = x - s$.) Because each change is, independently, a positive change with the same probability p, it follows, conditional on there being a total of $\frac{t}{2\Delta} + \frac{x-s}{2c}$ positive changes out of the first t/Δ changes, that all possible choices of the changes that were positive are equally likely. (That is, if a coin having probability p is flipped m times, then, given that k heads resulted, the subset of trials that resulted in heads is equally likely to be any of the $\binom{m}{k}$ subsets of size k.) Thus, even though p depends on μ, the conditional distribution of the history of prices up to time t, given that $X(t) = x$, does not depend on μ. (It does, however, depend on σ because c, the size of a change, depends on σ, and so if σ changed, then so would the number of the t/Δ changes that would have had to be positive for $S(t)$ to equal x.) Letting Δ go to 0 now completes the proof.   □

The Brownian motion process has a distinguished scientific pedigree.

It is named after the English botanist Robert Brown, who first described (in 1827) the unusual motion exhibited by a small particle that is totally immersed in a liquid or gas. The first explanation of this motion was given by Albert Einstein in 1905. He showed mathematically that Brownian motion could be explained by assuming that the immersed particle was continually being subjected to bombardment by the molecules of the surrounding medium. A mathematically concise definition, as well as an elucidation of some of the mathematical properties of Brownian motion, was given by the American applied mathematician Norbert Wiener in a series of papers originating in 1918.

Interestingly, Brownian motion was independently introduced in 1900 by the French mathematician Bachelier, who used it in his doctoral dissertation to model the price movements of stocks and commodities. However, Brownian motion appears to have two major flaws when used to model stock or commodity prices. First, since the price of a stock is a normal random variable, it can theoretically become negative. Second, the assumption that a price *difference* over an interval of fixed length has the same normal distribution no matter what the price at the beginning of the interval does not seem totally reasonable. For instance, many people might not think that the probability a stock presently selling at $20 would drop to $15 (a loss of 25%) in one month would be the same as the probability that when the stock is at $10 it would drop to $5 (a loss of 50%) in one month.

A process often used to model the price of a security as it evolves over time is the geometric Brownian motion process.

3.3   Geometric Brownian Motion

Definition   Let $X(t), t \geq 0$ be a Brownian motion process with drift parameter μ and variance parameter σ^2, and let

$$S(t) = e^{X(t)}, \quad t \geq 0$$

The process $S(t)$, $t \geq 0$, is said to be be a *geometric Brownian motion* process with drift parameter μ and variance parameter σ^2.

Let $S(t), t \geq 0$ be a geometric Brownian motion process with drift parameter μ and variance parameter σ^2. Because $\log(S(t))$, $t \geq 0$, is

Brownian motion and $\log(S(t + y)) - \log(S(y)) = \log(\frac{S(t+y)}{S(y)})$, it follows from the Brownian motion definition that for all positive y and t,

$$\log\left(\frac{S(t+y)}{S(y)}\right)$$

is independent of the process values up to time y and has a normal distribution with mean μt and variance $t\sigma^2$.

When used to model the price of a security over time, the geometric Brownian motion process possesses neither of the flaws of the Brownian motion process. Because it is the logarithm of the stock's price that is assumed to be normal random variable, the model does not allow for negative stock prices. Furthermore, because it is ratios, rather than differences, of prices separated by a fixed amount of time that have the same distribution, the geometric Brownian motion makes what many feel is the more reasonable assumption that it is the *percentage*, rather than the absolute, change in price whose probabilities do not depend on the current price.

Remarks:

• When geometric Brownian motion is used to model the price of a security over time, it is common to call σ the *volatility* parameter.
• If $S(0) = s$, then we can write

$$S(t) = se^{X(t)}, \quad t \geq 0$$

where $X(t)$, $t \geq 0$, is a Brownian motion process with $X(0) = 0$.
• If X is a normal random variable, then it can be shown that

$$E[e^X] = \exp\{E[X] + \text{Var}(X)/2\}$$

Hence, if $S(t), t \geq 0$, is a geometric Brownian motion process with drift μ and volatility σ having $S(0) = s$, then

$$E[S(t)] = se^{\mu t + t\sigma^2/2} = se^{(\mu + \sigma^2/2)t}$$

Thus, under geometric Brownian motion, the expected price of a security grows at rate $\mu + \sigma^2/2$. As a result, $\mu + \sigma^2/2$ is often called the *rate* of the geometric Brownian motion. Consequently, a geometric Brownian motion with rate parameter μ_r and volatility σ would have drift parameter $\mu_r - \sigma^2/2$.

3.3.1   *Geometric Brownian Motion as a Limit of Simpler Models*

Let $S(t), t \geq 0$ be a geometric Brownian motion process with drift parameter μ and volatility parameter σ. Because $X(t) = \log(S(t))$, $t \geq 0$, is Brownian motion, we can use its approximating process to obtain an approximating process for geometric Brownian motion. Using that $\frac{S(y+\Delta)}{S(y)} = e^{X(y+\Delta)-X(y)}$, we see that

$$S(y + \Delta) = S(y)e^{X(y+\Delta)-X(y)}$$

From the preceding it follows that we can approximate geometric Brownian motion by a model for the price of a security in which price changes occur only at times that are integral multiples of Δ. Moreover, whenever a change occurs, it results in the price of the security being multiplied either by the factor u with probability p or by the factor d with probability $1 - p$, where

$$u = e^{\sigma\sqrt{\Delta}}, \qquad d = e^{-\sigma\sqrt{\Delta}}$$

and

$$p = \frac{1}{2}\left(1 + \frac{\mu}{\sigma}\sqrt{\Delta}\right)$$

As Δ goes to 0, the preceding model becomes geometric Brownian motion. Consequently, geometric Brownian motion can be approximated by a relatively simple process that goes either up or down by fixed factors at regularly spaced times.

3.4   *The Maximum Variable

Let $X(v), v \geq 0$, be a Brownian motion process with drift parameter μ and variance parameter σ^2. Suppose that $X(0) = 0$, so that the process starts at state 0. Now, define

$$M(t) = \max_{0 \leq v \leq t} X(v)$$

to be the maximal value of the Brownian motion up to time t. In this section we derive first the conditional distribution of $M(t)$ given the value of $X(t)$ and then use this to derive the unconditional distribution of $M(t)$.

Theorem 3.4.1 *For y > x*

$$P(M(t) \geq y|X(t) = x) = e^{-2y(y-x)/t\sigma^2}, \quad y \geq 0$$

Proof. Because $X(0) = 0$, it follows that $M(t) \geq 0$, and so the result is true when $y = 0$ (as both sides are equal to 1 in this case). So suppose that $y > 0$. First note that it follows from Theorem 3.1.1 that $P(M(t) \geq y|X(t) = x)$ does not depend on the value of μ. So let us take $\mu = 0$. Now, let T_y denote the first time that the Brownian motion reaches the value y, and note that it follows from the continuity property of Brownian motion that the event that $M(t) \geq y$ is equivalent to the event that $T_y \leq t$. (This is true because before the process can exceed the positive value y it must, by continuity, first pass through that value.) Let h be a small positive number for which $y > x + h$. Then

$$
\begin{aligned}
&P(M(t) \geq y, \ x \leq X(t) \leq x + h) \\
&= P(T_y \leq t, \ x \leq X(t) \leq x + h) \\
&= P(x \leq X(t) \leq x + h|T_y \leq t)P(T_y \leq t) \quad (3.1)
\end{aligned}
$$

Now, given $T_y \leq t$, the event $x \leq X(t) \leq x + h$ will occur if, after hitting y, the additional amount $X(t) - X(T_y) = X(t) - y$ by which the process changes by time t is between $x - y$ and $x + h - y$. Because the distribution of this additional change is symmetric about 0 (since $\mu = 0$ and the distribution of a normal random variable is symmetric about its mean), it follows that the additional change is just as likely to be between $-(x + h - y)$ and $-(x - y)$ as it is to be between $x - y$ and $x + h - y$. Consequently,

$$
\begin{aligned}
&P(x \leq X(t) \leq x + h|T_y \leq t) \\
&= P(x - y \leq X(t) - y \leq x + h - y|T_y \leq t) \\
&= P(-(x + h - y) \leq X(t) - y \leq -(x - y)|T_y \leq t)
\end{aligned}
$$

The preceding, in conjunction with Equation (3.1), gives

$$
\begin{aligned}
&P(M(t) \geq y, \ x \leq X(t) \leq x + h) \\
&= P(2y - x - h \leq X(t) \leq 2y - x|T_y \leq t)P(T_y \leq t) \\
&= P(2y - x - h \leq X(t) \leq 2y - x, \ T_y \leq t) \\
&= P(2y - x - h \leq X(t) \leq 2y - x)
\end{aligned}
$$

The final equation following because the assumption $y > x + h$ yields that $2y - x - h > y$, and so, by the continuity of Brownian motion, $2y - x - h \leq X(t)$ implies that $T_y \leq t$. Hence,

$$P(M(t) \geq y | x \leq X(t) \leq x + h) = \frac{P(2y - x - h \leq X(t) \leq 2y - x)}{P(x \leq X(t) \leq x + h)}$$

$$\approx \frac{f_{X(t)}(2y - x)\, h}{f_{X(t)}(x)\, h} \quad \text{(for } h \text{ small)}$$

where $f_{X(t)}$, the density function of $X(t)$, is the density of a normal random variable with mean 0 and variance $t\sigma^2$. On letting $h \to 0$ in the preceding, we obtain that

$$P(M(t) \geq y | X(t) = x) = \frac{f_{X(t)}(2y - x)}{f_{X(t)}(x)}$$

$$= \frac{e^{-(2y-x)^2/2t\sigma^2}}{e^{-x^2/2t\sigma^2}}$$

$$= e^{-2y(y-x)/t\sigma^2} \qquad \square$$

With Z being a standard normal distribution function, let

$$\bar{\Phi}(x) = 1 - \Phi(x) = P(Z > x)$$

We now have

Corollary 3.4.1 *For* $y \geq 0$

$$P(M(t) \geq y) = e^{2y\mu/\sigma^2} \bar{\Phi}\left(\frac{\mu t + y}{\sigma \sqrt{t}}\right) + \bar{\Phi}\left(\frac{y - \mu t}{\sigma \sqrt{t}}\right)$$

Proof. Conditioning on $X(t)$, and using Theorem 3.4.1 gives

$$P(M(t) \geq y) = \int_{-\infty}^{\infty} P(M(t) \geq y | X(t) = x) f_{X(t)}(x) dx$$

$$= \int_{-\infty}^{y} P(M(t) \geq y | X(t) = x) f_{X(t)}(x) dx$$

$$+ \int_{y}^{\infty} P(M(t) \geq y | X(t) = x) f_{X(t)}(x) dx$$

$$= \int_{-\infty}^{y} e^{-2y(y-x)/t\sigma^2} f_{X(t)}(x) dx + \int_{y}^{\infty} f_{X(t)}(x) dx$$

Using the fact that $f_{X(t)}$ is the density function of a normal random variable with mean μt and variance $t\sigma^2$, the proof is completed by simplifying the right side of the preceding:

$$
P(M(t) \geq y)
$$

$$
= \int_{-\infty}^{y} e^{-2y(y-x)/t\sigma^2} \frac{1}{\sqrt{2\pi t\sigma^2}} e^{-(x-\mu t)^2/2t\sigma^2} \, dx + P(X(t) > y)
$$

$$
= \frac{1}{\sqrt{2\pi t} \, \sigma} e^{-2y^2/t\sigma^2} e^{-\mu^2 t^2/2t\sigma^2}
$$

$$
\times \int_{-\infty}^{y} \exp\left\{ -\frac{1}{2t\sigma^2} \left(x^2 - 2\mu t x - 4yx \right) \right\} dx + P(X(t) > y)
$$

$$
= \frac{1}{\sqrt{2\pi t} \, \sigma} e^{-(4y^2+\mu^2 t^2)/2t\sigma^2}
$$

$$
\times \int_{-\infty}^{y} \exp\left\{ -\frac{1}{2t\sigma^2} \left(x^2 - 2x(\mu t + 2y) \right) \right\} dx + P(X(t) > y)
$$

Now,

$$
x^2 - 2x(\mu t + 2y) = (x - (\mu t + 2y))^2 - (\mu t + 2y)^2
$$

giving that

$$
P(M(t) \geq y) = e^{-(4y^2+\mu^2 t^2-(\mu t+2y)^2)/2t\sigma^2} \frac{1}{\sqrt{2\pi t} \, \sigma}
$$

$$
\times \int_{-\infty}^{y} e^{-(x-\mu t-2y)^2/2t\sigma^2} \, dx + P(X(t) > y)
$$

Letting Z be a standard normal random variable, we obtain on making the change of variable

$$
w = \frac{x - \mu t - 2y}{\sigma\sqrt{t}}, \qquad dx = \sigma\sqrt{t}\, dw
$$

$$
P(M(t) \geq y) = e^{2y\mu/\sigma^2} \frac{1}{\sqrt{2\pi}} \int_{-\infty}^{\frac{-\mu t - y}{\sigma\sqrt{t}}} e^{-w^2/2} \, dw
$$

$$
+ P\left(\frac{X(t) - \mu t}{\sigma\sqrt{t}} > \frac{y - \mu t}{\sigma\sqrt{t}} \right)
$$

$$
= e^{2y\mu/\sigma^2} P\left(Z < \frac{-\mu t - y}{\sigma\sqrt{t}} \right) + P\left(Z > \frac{y - \mu t}{\sigma\sqrt{t}} \right)
$$

$$
= e^{2y\mu/\sigma^2} P\left(Z > \frac{\mu t + y}{\sigma\sqrt{t}} \right) + P\left(Z > \frac{y - \mu t}{\sigma\sqrt{t}} \right)
$$

and the proof is complete.   □

In the proof of Theorem 3.4.1 we let T_y denote the first time the Brownian motion is equal to y. That is,

$$T_y = \begin{cases} \infty, & \text{if } X(t) \neq y \text{ for all } t \geq 0 \\ \min(t : X(t) = y), & \text{otherwise} \end{cases}$$

In addition, as previously noted, it follows from the continuity of Brownian motion paths that, for $y > 0$, the process would have hit y by time t if and only if the maximum of the process by time t is at least y. That is,

$$T_y \leq t \Leftrightarrow M(t) \geq y$$

Hence, Corollary 3.4.1 yields that

$$P(T_y \leq t) = e^{2y\mu/\sigma^2}\bar{\Phi}\left(\frac{y + \mu t}{\sigma\sqrt{t}}\right) + \bar{\Phi}\left(\frac{y - \mu t}{\sigma\sqrt{t}}\right)$$

If we let $M_{\mu,\sigma}(t)$ denote a random variable having the distribution of the maximum value up to time t of a Brownian motion process that starts at 0 and has drift parameter μ and variance parameter σ^2, then the distribution of $M_{\mu,\sigma}(t)$ is given by Corollary 3.4.1. Now suppose we want the distribution of

$$M^*(t) = \min_{0 \leq v \leq t} X(v)$$

Using that $-X(v)$, $v \geq 0$, is a Brownian motion process with drift parameter $-\mu$ and variance parameter σ^2, we obtain for $y > 0$

$$\begin{aligned} P(M^*(t) \leq -y) &= P(\min_{0 \leq v \leq t} X(v) \leq -y) \\ &= P(-\max_{0 \leq v \leq t} -X(v) \leq -y) \\ &= P(\max_{0 \leq v \leq t} -X(v) \geq y) \\ &= P(M_{-\mu,\sigma}(t) \geq y) \\ &= e^{-2y\mu/\sigma^2}\bar{\Phi}\left(\frac{-\mu t + y}{\sigma\sqrt{t}}\right) + \bar{\Phi}\left(\frac{y + \mu t}{\sigma\sqrt{t}}\right) \end{aligned}$$

where the final equality used Corollary 3.4.1.

3.5   The Cameron-Martin Theorem

For an underlying Brownian motion process with variance parameter σ^2, let us use the notation E_μ to denote that we are taking expectations under the assumption that the drift parameter is μ. Thus, for instance, E_0 would signify that the expectation is taken under the assumption that the drift parameter of the Brownian motion process is 0. The following is known as the Cameron-Martin theorem. (It is a special case of a more general result, known as Girsanov's theorem.)

Theorem 3.5.1 *Let W be a random variable whose value is determined by the history of the Brownian motion up to time t. That is, the value of W is determined by a knowledge of the values of $X(s)$, $0 \le s \le t$. Then,*

$$E_\mu[W] = e^{-\mu^2 t/2\sigma^2} E_0[W e^{\mu X(t)/\sigma^2}]$$

Proof. Conditioning on $X(t)$, which is normal with mean μt and variance $t\sigma^2$, yields

$$E_\mu[W] = \int_{-\infty}^{\infty} E_\mu[W|X(t) = x] \frac{1}{\sqrt{2\pi t\sigma^2}} e^{-(x-\mu t)^2/2t\sigma^2} \, dx$$

$$= \int_{-\infty}^{\infty} E_0[W|X(t) = x] \frac{1}{\sqrt{2\pi t\sigma^2}} e^{-(x-\mu t)^2/2t\sigma^2} \, dx$$

$$= \int_{-\infty}^{\infty} E_0[W|X(t) = x] \frac{1}{\sqrt{2\pi t\sigma^2}} e^{-x^2/2t\sigma^2} e^{(2\mu x - \mu^2 t)/2\sigma^2} \, dx$$

$$(3.2)$$

where the second equality follows from Theorem 3.1.1, which states that, given $X(t) = x$, the conditional distribution of the process up to time t (and thus the conditional distribution of W) is the same for all values μ. Now, if we define

$$Y = e^{-\mu^2 t/2\sigma^2} e^{\mu X(t)/\sigma^2} = e^{(2\mu X(t) - \mu^2 t)/2\sigma^2}$$

then

$$E_0[WY] = \int_{-\infty}^{\infty} E_0[WY|X(t) = x] \frac{1}{\sqrt{2\pi t\sigma^2}} e^{-x^2/2t\sigma^2} \, dx$$

But, given that $X(t) = x$, the random variable Y is equal to the constant $e^{(2\mu x - \mu^2 t)/2\sigma^2}$, and so the preceding yields

$$E_0[WY] = \int_{-\infty}^{\infty} e^{(2\mu x - \mu^2 t)/2\sigma^2} E_0[W | X(t) = x] \frac{1}{\sqrt{2\pi t\sigma^2}} e^{-x^2/2t\sigma^2} \, dx$$
$$= E_\mu[W]$$

where the final equality used (3.2). $\square$

3.6   Exercises

Exercise 3.1 If $X(t), t \geq 0$ is a Brownian motion process with drift parameter μ and variance parameter σ^2 for which $X(0) = 0$, show that $-X(t), t \geq 0$ is a Brownian motion process with drift parameter $-\mu$ and variance parameter σ^2.

Exercise 3.2 Let $X(t), t \geq 0$ be a Brownian motion process with drift parameter $\mu = 3$ and variance parameter $\sigma^2 = 9$. If $X(0) = 10$, find

(a) $E[X(2)]$;
(b) $\text{Var}(X(2))$;
(c) $P(X(2) > 20)$;
(d) $P(X(.5) > 10)$.

Exercise 3.3 Let $\Delta = 0.1$ in the approximation model to the Brownian motion process of the preceding problem. For this approximation model, find

(a) $E[X(1)]$;
(b) $\text{Var}(X(1))$;
(c) $P(X(.5) > 10)$.

Exercise 3.4 Let $S(t), t \geq 0$ be a geometric Brownian motion process with drift parameter $\mu = 0.1$ and volatility parameter $\sigma = 0.2$. Find

(a) $P(S(1) > S(0))$;
(b) $P(S(2) > S(1) > S(0))$;
(c) $P(S(3) < S(1) > S(0))$.

Exercise 3.5 Repeat Exercise 3.4 when the volatility parameter is 0.4.

Exercise 3.6 Let $S(t)$, $t \geq 0$ be a geometric Brownian motion process with drift parameter μ and volatility parameter σ. Assuming that $S(0) = s$, find Var$(S(t))$. *Hint:* Use the identity

$$\text{Var}(X) = E[X^2] - (E[X])^2$$

Exercise 3.7 Let $\{X(t), t \geq 0\}$ be a Brownian motion process with drift parameter μ and variance parameter σ^2. Assume that $X(0) = 0$, and let T_y be the first time that the process is equal to y. For $y > 0$, show that

$$P(T_y < \infty) = \begin{cases} 1, & \text{if } \mu \geq 0 \\ e^{2y\mu/\sigma^2}, & \text{if } \mu < 0 \end{cases}$$

Let $M = \max_{0 < t < \infty} X(t)$ be the maximal value ever attained by the process, and conclude from the preceding that, when $\mu < 0$, M has an exponential distribution with rate $-2\mu/\sigma^2$.

Exercise 3.8 Let $S(v)$, $v \geq 0$ be a geometric Brownian motion process with drift parameter μ and volatility parameter σ, having $S(0) = s$. Find $P(\max_{0 \leq v \leq t} S(v) \geq y)$.

Exercise 3.9 Find $P(\max_{0 \leq v \leq 1} S(v) < 1.2\, S(0))$ when $S(v)$, $v \geq 0$, is geometric Brownian motion with drift .1 and volatility .3.

4. Interest Rates and Present Value Analysis

4.1 Interest Rates

If you borrow the amount P (called the principal), which must be repaid after a time T along with simple interest at rate r per time T, then the amount to be repaid at time T is

$$P + rP = P(1 + r).$$

That is, you must repay both the principal P and the interest, equal to the principal times the interest rate. For instance, if you borrow \$100 to be repaid after one year with a simple interest rate of 5% per year (i.e., $r = .05$), then you will have to repay \$105 at the end of the year.

Example 4.1a Suppose that you borrow the amount P, to be repaid after one year along with interest at a rate r per year *compounded* semiannually. What does this mean? How much is owed in a year?

Solution. In order to solve this example, you must realize that having your interest compounded semiannually means that after half a year you are to be charged simple interest at the rate of $r/2$ per half-year, and that interest is then added on to your principal, which is again charged interest at rate $r/2$ for the second half-year period. In other words, after six months you owe

$$P(1 + r/2).$$

This is then regarded as the new principal for another six-month loan at interest rate $r/2$; hence, at the end of the year you will owe

$$P(1 + r/2)(1 + r/2) = P(1 + r/2)^2. \qquad \square$$

Example 4.1b If you borrow \$1,000 for one year at an interest rate of 8% per year compounded quarterly, how much do you owe at the end of the year?

Solution. An interest rate of 8% that is compounded quarterly is equivalent to paying simple interest at 2% per quarter-year, with each successive quarter charging interest not only on the original principal but also on the interest that has accrued up to that point. Thus, after one quarter you owe

$$1,000(1 + .02);$$

after two quarters you owe

$$1,000(1 + .02)(1 + .02) = 1,000(1 + .02)^2;$$

after three quarters you owe

$$1,000(1 + .02)^2(1 + .02) = 1,000(1 + .02)^3;$$

and after four quarters you owe

$$1,000(1 + .02)^3(1 + .02) = 1,000(1 + .02)^4 = \$1,082.40. \qquad \square$$

Example 4.1c Many credit-card companies charge interest at a yearly rate of 18% compounded monthly. If the amount P is charged at the beginning of a year, how much is owed at the end of the year if no previous payments have been made?

Solution. Such a compounding is equivalent to paying simple interest every month at a rate of $18/12 = 1.5\%$ per month, with the accrued interest then added to the principal owed during the next month. Hence, after one year you will owe

$$P(1 + .015)^{12} = 1.1956P. \qquad \square$$

If the interest rate r is compounded then, as we have seen in Examples 4.1b and 4.1c, the amount of interest actually paid is greater than if we were paying simple interest at rate r. The reason, of course, is that in compounding we are being charged interest on the interest that has already been computed in previous compoundings. In these cases, we call r the *nominal* interest rate, and we define the *effective interest rate*, call it r_{eff}, by

$$r_{\text{eff}} = \frac{\text{amount repaid at the end of a year} - P}{P}.$$

For instance, if the loan is for one year at a nominal interest rate r that is to be compounded quarterly, then the effective interest rate for the year is

$$r_{\text{eff}} = (1 + r/4)^4 - 1.$$

Thus, in Example 4.1b the effective interest rate is 8.24% whereas in Example 4.1c it is 19.56%. Since

$$P(1 + r_{\text{eff}}) = \text{amount repaid at the end of a year},$$

the payment made in a one-year loan with compound interest is the same as if the loan called for simple interest at rate r_{eff} per year.

Example 4.1d *The Doubling Rule* If you put funds into an account that pays interest at rate r compounded annually, how many years does it take for your funds to double?

Solution. Since your initial deposit of D will be worth $D(1 + r)^n$ after n years, we need to find the value of n such that

$$(1 + r)^n = 2.$$

Now,

$$(1 + r)^n = \left(1 + \frac{nr}{n}\right)^n$$
$$\approx e^{nr},$$

where the approximation is fairly precise provided that n is not too small. Therefore,

$$e^{nr} \approx 2,$$

implying that

$$n \approx \frac{\log(2)}{r} = \frac{.693}{r}.$$

Thus, it will take n years for your funds to double when

$$n \approx \frac{.7}{r}.$$

For instance, if the interest rate is 1% ($r = .01$) then it will take approximately 70 years for your funds to double; if $r = .02$, it will take about

35 years; if $r = .03$, it will take about $23\frac{1}{3}$ years; if $r = .05$, it will take about 14 years; if $r = .07$, it will take about 10 years; and if $r = .10$, it will take about 7 years.

As a check on the preceding approximations, note that (to three–decimal-place accuracy):

$$(1.01)^{70} = 2.007,$$

$$(1.02)^{35} = 2.000,$$

$$(1.03)^{23.33} = 1.993,$$

$$(1.05)^{14} = 1.980,$$

$$(1.07)^{10} = 1.967,$$

$$(1.10)^{7} = 1.949. \qquad \square$$

Suppose now that we borrow the principal P for one year at a nominal interest rate of r per year, compounded *continuously*. Now, how much is owed at the end of the year? Of course, to answer this we must first decide on an appropriate definition of "continuous" compounding. To do so, note that if the loan is compounded at n equal intervals in the year, then the amount owed at the end of the year is $P(1 + r/n)^n$. As it is reasonable to suppose that continuous compounding refers to the limit of this process as n grows larger and larger, the amount owed at time 1 is

$$P \lim_{n \to \infty} (1 + r/n)^n = Pe^r.$$

Example 4.1e If a bank offers interest at a nominal rate of 5% compounded continuously, what is the effective interest rate per year?

Solution. The effective interest rate is

$$r_{\text{eff}} = \frac{Pe^{.05} - P}{P} = e^{.05} - 1 \approx .05127.$$

That is, the effective interest rate is 5.127% per year. $\qquad \square$

If the amount P is borrowed for t years at a nominal interest rate of r per year compounded continuously, then the amount owed at time t is Pe^{rt}. This follows because if interest is compounded n times during the year, then there would have been nt compoundings by time t, giving a debt level of $P(1 + r/n)^{nt}$. Consequently, under continuous compounding the debt at time t would be

$$P \lim_{n \to \infty} \left(1 + \frac{r}{n}\right)^{nt} = P\left(\lim_{n \to \infty}\left(1 + \frac{r}{n}\right)^n\right)^t$$

$$= Pe^{rt}.$$

It follows from the preceding that continuous compounded interest at rate r per unit time can be interpreted as being a continuous compounding of a nominal interest rate of rt per (unit of time) t.

4.2   Present Value Analysis

Suppose that one can both borrow and loan money at a nominal rate r per period that is compounded periodically. Under these conditions, what is the present worth of a payment of v dollars that will be made at the end of period i? Since a bank loan of $v(1 + r)^{-i}$ would require a payoff of v at period i, it follows that the *present value* of a payoff of v to be made at time period i is $v(1 + r)^{-i}$.

The concept of present value enables us to compare different income streams to see which is preferable.

Example 4.2a Suppose that you are to receive payments (in thousands of dollars) at the end of each of the next five years. Which of the following three payment sequences is preferable?

A. 12, 14, 16, 18, 20;
B. 16, 16, 15, 15, 15;
C. 20, 16, 14, 12, 10.

Solution. If the nominal interest rate is r compounded yearly, then the present value of the sequence of payments x_i ($i = 1, 2, 3, 4, 5$) is

$$\sum_{i=1}^{5}(1 + r)^{-i}x_i;$$

the sequence having the largest present value is preferred. It thus follows that the superior sequence of payments depends on the interest rate.

Table 4.1: *Present Values*

r	A	B	C
.1	59.21	58.60	56.33
.2	45.70	46.39	45.69
.3	36.49	37.89	38.12

Payment Sequence (header spanning A, B, C)

If r is small, then the sequence **A** is best since its sum of payments is the highest. For a somewhat larger value of r, the sequence **B** would be best because – although the total of its payments (77) is less than that of **A** (80) – its earlier payments are larger than are those of **A**. For an even larger value of r, the sequence **C**, whose earlier payments are higher than those of either **A** or **B**, would be best. Table 4.1 gives the present values of these payment streams for three different values of r.

It should be noted that the payment sequences can be compared according to their values at any specified time. For instance, to compare them in terms of their time-5 values, we would determine which sequence of payments yields the largest value of

$$\sum_{i=1}^{5}(1+r)^{5-i}x_i = (1+r)^5 \sum_{i=1}^{5}(1+r)^{-i}x_i.$$

Consequently, we obtain the same preference ordering as a function of interest rate as before. □

Remark. Let the given interest rate be r, compounded yearly. Any cash flow stream $\mathbf{a} = a_1, a_2, \ldots, a_n$ that returns you a_i dollars at the end of year i (for each $i = 1, \ldots, n$) can be replicated by depositing

$$PV(\mathbf{a}) = \frac{a_1}{1+r} + \frac{a_2}{(1+r)^2} + \cdots + \frac{a_n}{(1+r)^n}$$

in a bank at time 0 and then making the successive withdrawals $a_1, a_2, \ldots, a_n$. To verify this claim, note that withdrawing a_1 at the end of year 1 would leave you with

$$(1+r)\left[\frac{a_1}{1+r} + \frac{a_2}{(1+r)^2} + \cdots + \frac{a_n}{(1+r)^n}\right] - a_1$$

$$= \frac{a_2}{(1+r)} + \cdots + \frac{a_n}{(1+r)^{n-1}}$$

on deposit. Thus, after withdrawing a_2 at the end of year 2 you would have

$$(1+r)\left[\frac{a_2}{1+r} + \cdots + \frac{a_n}{(1+r)^{n-1}}\right] - a_2 = \frac{a_3}{(1+r)} + \cdots + \frac{a_n}{(1+r)^{n-2}}.$$

Continuing, it follows that withdrawing a_i at the end of year i ($i < n$) would leave you with

$$\frac{a_{i+1}}{(1+r)} + \cdots + \frac{a_n}{(1+r)^{n-i}}$$

on deposit. Consequently, you would have $a_n/(1+r)$ on deposit after withdrawing a_{n-1}, and this is just enough to cover your next withdrawal of a_n at the end of the following year.

In a similar manner, the cash flow sequence $a_1, a_2, \ldots, a_n$ can be transformed into the initial capital PV(**a**) by borrowing this amount from a bank and then using the cash flow to pay off this debt. Therefore, any cash flow sequence is equivalent to an initial reception of the present value of the cash flow sequence, thus showing that one cash flow sequence is preferable to another whenever the former has a larger present value than the latter.   □

Example 4.2b   A company needs a certain type of machine for the next five years. They presently own such a machine, which is now worth $6,000 but will lose $2,000 in value in each of the next three years, after which it will be worthless and unuseable. The (beginning-of-the-year) value of its yearly operating cost is $9,000, with this amount expected to increase by $2,000 in each subsequent year that it is used. A new machine can be purchased at the beginning of any year for a fixed cost of $22,000. The lifetime of a new machine is six years, and its value decreases by $3,000 in each of its first two years of use and then by $4,000 in each following year. The operating cost of a new machine is $6,000 in its first year, with an increase of $1,000 in each subsequent year. If the interest rate is 10%, when should the company purchase a new machine?

Solution. The company can purchase a new machine at the beginning of year 1, 2, 3, or 4, with the following six-year cash flows (in units of $1,000) as a result:

• buy at beginning of year 1: 22, 7, 8, 9, 10, −4;
• buy at beginning of year 2: 9, 24, 7, 8, 9, −8;
• buy at beginning of year 3: 9, 11, 26, 7, 8, −12;
• buy at beginning of year 4: 9, 11, 13, 28, 7, −16.

To see why this listing is correct, suppose that the company will buy a new machine at the beginning of year 3. Then its year-1 cost is the $9,000 operating cost of the old machine; its year-2 cost is the $11,000 operating cost of this machine; its year-3 cost is the $22,000 cost of a new machine, plus the $6,000 operating cost of this machine, minus the $2,000 obtained for the replaced machine; its year-4 cost is the $7,000 operating cost; its year-5 cost is the $8,000 operating cost; and its year-6 cost is −$12, 000, the negative of the value of the 3-year-old machine that it no longer needs. The other cash flow sequences are similarly argued.

With the yearly interest rate $r = .10$, the present value of the first cost-flow sequence is

$$22 + \frac{7}{1.1} + \frac{8}{(1.1)^2} + \frac{9}{(1.1)^3} + \frac{10}{(1.1)^4} - \frac{4}{(1.1)^5} = 46.083.$$

The present values of the other cash flows are similarly determined, and the four present values are

$$46.083, \quad 43.794, \quad 43.760, \quad 45.627.$$

Therefore, the company should purchase a new machine two years from now. □

Example 4.2c An individual who plans to retire in 20 years has decided to put an amount A in the bank at the beginning of each of the next 240 months, after which she will withdraw $1,000 at the beginning of each of the following 360 months. Assuming a nominal yearly interest rate of of 6% compounded monthly, how large does A need to be?

Solution. Let $r = .06/12 = .005$ be the monthly interest rate. With $\beta = \frac{1}{1+r}$, the present value of all her deposits is

$$A + A\beta + A\beta^2 + \cdots + A\beta^{239} = A\frac{1 - \beta^{240}}{1 - \beta}.$$

Similarly, if W is the amount withdrawn in the following 360 months, then the present value of all these withdrawals is

$$W\beta^{240} + W\beta^{241} + \cdots + W\beta^{599} = W\beta^{240}\frac{1 - \beta^{360}}{1 - \beta}.$$

Thus she will be able to fund all withdrawals (and have no money left in her account) if

$$A\frac{1 - \beta^{240}}{1 - \beta} = W\beta^{240}\frac{1 - \beta^{360}}{1 - \beta}.$$

With $W = 1,000$, and $\beta = 1/1.005$, this gives

$$A = 360.99.$$

That is, saving $361 a month for 240 months will enable her to withdraw $1,000 a month for the succeeding 360 months.

Remark. In this example we have made use of the algebraic identity

$$1 + b + b^2 + \cdots + b^n = \frac{1 - b^{n+1}}{1 - b}.$$

We can prove this identity by letting

$$x = 1 + b + b^2 + \cdots + b^n$$

and then noting that

$$x - 1 = b + b^2 + \cdots + b^n$$
$$= b(1 + b + \cdots + b^{n-1})$$
$$= b(x - b^n).$$

Therefore,

$$(1 - b)x = 1 - b^{n+1},$$

which yields the identity.

It can be shown by the same technique, or by letting n go to infinity, that when $|b| < 1$ we have

$$1 + b + b^2 + \cdots = \frac{1}{1-b}.$$ □

Example 4.2d A perpetuity entitles its holder to be paid the constant amount c at the end of each of an infinite sequence of years. That is, it pays its holder c at the end of year i for each $i = 1, 2, \ldots$. If the interest rate is r, compounded yearly, then what is the present value of such a cash flow sequence?

Solution. Because such a cash flow could be replicated by initially putting the principle c/r in the bank and then withdrawing the interest earned (leaving the principal intact) at the end of each period, whereas it could not be replicated by putting any smaller amount in the bank, it would seem that the present value of the infinite flow is c/r. This intuition is easily checked mathematically by

$$
\begin{aligned}
\text{PV} &= \frac{c}{1+r} + \frac{c}{(1+r)^2} + \frac{c}{(1+r)^3} + \cdots \\
&= \frac{c}{1+r}\left[1 + \frac{1}{1+r} + \frac{1}{(1+r)^2} + \cdots \right] \\
&= \frac{c}{1+r}\frac{1}{1 - \frac{1}{1+r}} \\
&= \frac{c}{r}.
\end{aligned}
$$ □

Example 4.2e Suppose you have just spoken to a bank about borrowing $100,000 to purchase a house, and the loan officer has told you that a $100,000 loan, to be repaid in monthly installments over 15 years with an interest rate of .6% per month, could be arranged. If the bank charges a loan initiation fee of $600, a house inspection fee of $400, and 1 "point," what is the effective annual interest rate of the loan being offered?

Solution. To begin, let us determine the monthly mortgage payment, call it A, of such a loan. Since $100,000 is to be repaid in 180 monthly payments at an interest rate of .6% per month, it follows that

$$A[\alpha + \alpha^2 + \cdots + \alpha^{180}] = 100,000,$$

where $\alpha = 1/1.006$. Therefore,

$$A = \frac{100,000(1 - \alpha)}{\alpha(1 - \alpha^{180})} = 910.05.$$

So if you were actually receiving $100,000 to be repaid in 180 monthly payments of $910.05, then the effective monthly interest rate would be .6%. However, taking into account the initiation and inspection fees involved and the bank charge of 1 point (which means that 1% of the nominal loan of $100,000 must be paid to the bank when the loan is received), it follows that you are actually receiving only $98,000. Consequently, the effective monthly interest rate is that value of r such that

$$A[\beta + \beta^2 + \cdots + \beta^{180}] = 98,000,$$

where $\beta = (1 + r)^{-1}$. Therefore,

$$\frac{\beta(1 - \beta^{180})}{1 - \beta} = 107.69$$

or, since $\frac{1-\beta}{\beta} = r$,

$$\frac{1 - \left(\frac{1}{1+r}\right)^{180}}{r} = 107.69.$$

Numerically solving this by trial and error (easily accomplished since we know that $r > .006$) yields the solution

$$r = .00627.$$

Since $(1 + .00627)^{12} = 1.0779$, it follows that what was quoted as a monthly interest rate of .6% is, in reality, an effective annual interest rate of approximately 7.8%.   □

Example 4.2f Suppose that one takes a mortgage loan for the amount L that is to be paid back over n months with equal payments of A at the end of each month. The interest rate for the loan is r per month, compounded monthly.

(a) In terms of L, n, and r, what is the value of A?
(b) After payment has been made at the end of month j, how much additional loan principal remains?

(c) How much of the payment during month j is for interest and how much is for principal reduction? (This is important because some contracts allow for the loan to be paid back early and because the interest part of the payment is tax-deductible.)

Solution. The present value of the n monthly payments is

$$\frac{A}{1+r} + \frac{A}{(1+r)^2} + \cdots + \frac{A}{(1+r)^n} = \frac{A}{1+r}\frac{1 - \left(\frac{1}{1+r}\right)^n}{1 - \frac{1}{1+r}}$$

$$= \frac{A}{r}[1 - (1+r)^{-n}].$$

Since this must equal the loan amount L, we see that

$$A = \frac{Lr}{1 - (1+r)^{-n}} = \frac{L(\alpha - 1)\alpha^n}{\alpha^n - 1}, \tag{4.1}$$

where

$$\alpha = 1 + r.$$

For instance, if the loan is for \$100,000 to be paid back over 360 months at a nominal yearly interest rate of .09 compounded monthly, then $r = .09/12 = .0075$ and the monthly payment (in dollars) would be

$$A = \frac{100,000(.0075)(1.0075)^{360}}{(1.0075)^{360} - 1} = 804.62.$$

Let R_j denote the remaining amount of principal owed after the payment at the end of month j ($j = 0, \ldots, n$). To determine these quantities, note that if one owes R_j at the end of month j then the amount owed immediately before the payment at the end of month $j + 1$ is $(1+r)R_j$; because one then pays the amount A, it follows that

$$R_{j+1} = (1+r)R_j - A = \alpha R_j - A.$$

Starting with $R_0 = L$, we obtain:

$$R_1 = \alpha L - A;$$

$$R_2 = \alpha R_1 - A$$

$$= \alpha(\alpha L - A) - A$$

$$= \alpha^2 L - (1 + \alpha)A;$$

$$R_3 = \alpha R_2 - A$$

$$= \alpha(\alpha^2 L - (1+\alpha)A) - A$$

$$= \alpha^3 L - (1+\alpha+\alpha^2)A.$$

In general, for $j = 0, \ldots, n$ we obtain

$$R_j = \alpha^j L - A(1 + \alpha + \cdots + \alpha^{j-1})$$

$$= \alpha^j L - A\frac{\alpha^j - 1}{\alpha - 1}$$

$$= \alpha^j L - \frac{L\alpha^n(\alpha^j - 1)}{\alpha^n - 1} \qquad \text{(from (4.1))}$$

$$= \frac{L(\alpha^n - \alpha^j)}{\alpha^n - 1}.$$

Let I_j and P_j denote the amounts of the payment at the end of month j that are for interest and for principal reduction, respectively. Then, since R_{j-1} was owed at the end of the previous month, we have

$$I_j = rR_{j-1}$$

$$= \frac{L(\alpha - 1)(\alpha^n - \alpha^{j-1})}{\alpha^n - 1}$$

and

$$P_j = A - I_j$$

$$= \frac{L(\alpha - 1)}{\alpha^n - 1}[\alpha^n - (\alpha^n - \alpha^{j-1})]$$

$$= \frac{L(\alpha - 1)\alpha^{j-1}}{\alpha^n - 1}.$$

As a check, note that

$$\sum_{j=1}^{n} P_j = L.$$

It follows that the amount of principal repaid in succeeding months increases by the factor $\alpha = 1 + r$. For example, in a \$100,000 loan for 30 years at a nominal interest rate of 9% per year compounded monthly, only \$54.62 of the \$804.62 paid during the first month goes toward reducing the principal of the loan; the remainder is interest. In each succeeding month, the amount of the payment that goes toward the principal increases by the factor 1.0075. □

Consider two cash flow sequences,

$$b_1, b_2, \ldots, b_n \quad \text{and} \quad c_1, c_2, \ldots, c_n.$$

Under what conditions is the present value of the first sequence at least as large as that of the second for every positive interest rate r? Clearly, $b_i \geq c_i$ $(i = 1, \ldots, n)$ is a sufficient condition. However, we can obtain weaker sufficient conditions. Let

$$B_i = \sum_{j=1}^{i} b_j \quad \text{and} \quad C_i = \sum_{j=1}^{i} c_j \quad \text{for } i = 1, \ldots, n;$$

then it can be shown that the condition

$$B_i \geq C_i \quad \text{for each } i = 1, \ldots, n$$

suffices. An even weaker sufficient condition is given by the following proposition.

Proposition 4.2.1 *If $B_n \geq C_n$ and if*

$$\sum_{i=1}^{k} B_i \geq \sum_{i=1}^{k} C_i$$

for each $k = 1, \ldots, n$, then

$$\sum_{i=1}^{n} b_i(1 + r)^{-i} \geq \sum_{i=1}^{n} c_i(1 + r)^{-i}$$

for every $r > 0$.

In other words, Proposition 4.2.1 states that the cash flow sequence $b_1, \ldots, b_n$ will, for every positive interest rate r, have a larger present value than the cash flow sequence $c_1, \ldots, c_n$ if (i) the total of the b cash flows is at least as large as the total of the c cash flows and (ii) for every $k = 1, \ldots, n$,

$$kb_1 + (k-1)b_2 + \cdots + b_k \geq kc_1 + (k-1)c_2 + \cdots + c_k.$$

4.3   Rate of Return

Consider an investment that, for an initial payment of a ($a > 0$), returns the amount b after one period. The *rate of return* on this investment is defined to be the interest rate r that makes the present value of the return equal to the initial payment. That is, the rate of return is that value r such that

$$\frac{b}{1+r} = a \quad \text{or} \quad r = \frac{b}{a} - 1.$$

Thus, for example, a \$100 investment that returns \$150 after one year is said to have a yearly rate of return of .50.

More generally, consider an investment that, for an initial payment of a ($a > 0$), yields a string of nonnegative returns $b_1, \ldots, b_n$. Here b_i is to be received at the end of period i ($i = 1, \ldots, n$), and $b_n > 0$. We define the rate of return per period of this investment to be the value of the interest rate such that the present value of the cash flow sequence is equal to zero when values are compounded periodically at that interest rate. That is, if we define the function P by

$$P(r) = -a + \sum_{i=1}^{n} b_i (1+r)^{-i}, \tag{4.2}$$

then the rate of return per period of the investment is that value $r^* > -1$ for which

$$P(r^*) = 0.$$

It follows from the assumptions $a > 0$, $b_i \geq 0$, and $b_n > 0$ that $P(r)$ is a strictly decreasing function of r when $r > -1$, implying (since $\lim_{r \to -1} P(r) = \infty$ and $\lim_{r \to \infty} P(r) = -a < 0$) that there is a unique value r^* satisfying the preceding equation. Moreover, since

$$P(0) = \sum_{i=1}^{n} b_i - a,$$

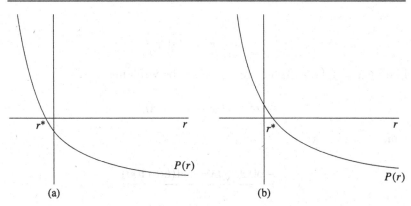

Figure 4.1: $P(r) = -a + \sum_{i \geq 1} b_i (1 + r)^{-i}$: (a) $\sum_i b_i < a$; (b) $\sum_i b_i > a$

it follows (see Figure 4.1) that r^* will be positive if

$$\sum_{i=1}^{n} b_i > a$$

and that r^* will be negative if

$$\sum_{i=1}^{n} b_i < a.$$

That is, there is a positive rate of return if the total of the amounts received exceeds the initial investment, and there is a negative rate of return if the reverse holds. Moreover, because of the monotonicity of $P(r)$, it follows that the cash flow sequence will have a positive present value when the interest rate is less than r^* and a negative present value when the interest rate is greater than r^*.

When an investment's rate of return is r^* per period, we often say that the investment yields a $100r^*$ percent rate of return per period.

Example 4.3a Find the rate of return from an investment that, for an initial payment of 100, yields returns of 60 at the end of each of the first two periods.

Solution. The rate of return will be the solution to

$$100 = \frac{60}{1+r} + \frac{60}{(1+r)^2}.$$

Letting $x = 1/(1+r)$, the preceding can be written as

$$60x^2 + 60x - 100 = 0,$$

which yields that

$$x = \frac{-60 \pm \sqrt{60^2 + 4(60)(100)}}{120}.$$

Since $-1 < r$ implies that $x > 0$, we obtain the solution

$$x = \frac{\sqrt{27,600} - 60}{120} \approx .8844.$$

Hence, the rate of return r^* is such that

$$1 + r^* \approx \frac{1}{.8844} \approx 1.131.$$

That is, the investment yields a rate of return of approximately 13.1% per period. ☐

The rate of return of investments whose string of payments spans more than two periods will usually have to be numerically determined. Because of the monotonicity of $P(r)$, a trial-and-error approach is usually quite efficient.

Remarks. (1) If we interpret the cash flow sequence by supposing that $b_1, \ldots, b_n$ represent the successive periodic payments made to a lender who loans a to a borrower, then the lender's periodic rate of return r^* is exactly the effective interest rate per period paid by the borrower.

(2) The quantity r^* is also sometimes called the *internal rate of return*.

Consider now a more general investment cash flow sequence $c_0, c_1, \ldots, c_n$. Here, if $c_i \geq 0$ then the amount c_i is received by the investor at the end of period i, and if $c_i < 0$ then the amount $-c_i$ must be paid by the investor at the end of period i. If we let

$$P(r) = \sum_{i=0}^{n} c_i(1 + r)^{-i}$$

be the present value of this cash flow when the interest rate is r per period, then in general there will not necessarily be a unique solution of the equation

$$P(r) = 0$$

in the region $r > -1$. As a result, the rate-of-return concept is unclear in the case of more general cash flows than the ones considered here. In addition, even in cases where we can show that the preceding equation has a unique solution r^*, it may result that $P(r)$ is not a monotone function of r; consequently, we could *not* assert that the investment yields a positive present value return when the interest rate is on one side of r^* and a negative present value return when it is on the other side.

One general situation for which we can prove that there is a unique solution is when the cash flow sequence starts out negative (resp. positive), eventually becomes positive (negative), and then remains nonnegative (nonpositive) from that point on. In other words, the sequence $c_0, c_1, \ldots, c_n$ has a single sign change. It then follows – upon using Descartes' rule of sign, along with the known existence of at least one solution – that there is a unique solution of the equation $P(r) = 0$ in the region $r > -1$.

4.4   Continuously Varying Interest Rates

Suppose that interest is continuously compounded but with a rate that is changing in time. Let the present time be time 0, and let $r(s)$ denote the interest rate at time s. Thus, if you put x in a bank at time s, then the

amount in your account at time $s + h \approx x(1 + r(s)h)$ (h small).

The quantity $r(s)$ is called the *spot* or the *instantaneous* interest rate at time s.

Let $D(t)$ be the amount that you will have on account at time t if you deposit 1 at time 0. In order to determine $D(t)$ in terms of the interest rates $r(s)$, $0 \le s \le t$, note that (for h small) we have

$$D(s + h) \approx D(s)(1 + r(s)h)$$

or

$$D(s + h) - D(s) \approx D(s)r(s)h$$

or

$$\frac{D(s + h) - D(s)}{h} \approx D(s)r(s).$$

The preceding approximation becomes exact as h becomes smaller and smaller. Hence, taking the limit as $h \to 0$, it follows that

$$D'(s) = D(s)r(s)$$

or

$$\frac{D'(s)}{D(s)} = r(s),$$

implying that

$$\int_0^t \frac{D'(s)}{D(s)}\, ds = \int_0^t r(s)\, ds$$

or

$$\log(D(t)) - \log(D(0)) = \int_0^t r(s)\, ds.$$

Since $D(0) = 1$, we obtain from the preceding equation that

$$D(t) = \exp\left\{ \int_0^t r(s)\, ds \right\}.$$

Now let $P(t)$ denote the present (i.e. time-0) value of the amount 1 that is to be received at time t ($P(t)$ would be the cost of a bond that yields a return of 1 at time t; it would equal e^{-rt} if the interest rate were always equal to r). Because a deposit of $1/D(t)$ at time 0 will be worth 1 at time t, we see that

$$P(t) = \frac{1}{D(t)} = \exp\left\{ -\int_0^t r(s)\, ds \right\}. \qquad (4.3)$$

Let $\bar{r}(t)$ denote the average of the spot interest rates up to time t; that is,

$$\bar{r}(t) = \frac{1}{t} \int_0^t r(s)\, ds.$$

The function $\bar{r}(t)$, $t \geq 0$, is called the *yield curve*.

Example 4.4a Find the yield curve and the present value function if

$$r(s) = \frac{1}{1+s}r_1 + \frac{s}{1+s}r_2.$$

Solution. Rewriting $r(s)$ as

$$r(s) = r_2 + \frac{r_1 - r_2}{1+s}, \quad s \geq 0,$$

shows that the yield curve is given by

$$\bar{r}(t) = \frac{1}{t} \int_0^t \left(r_2 + \frac{r_1 - r_2}{1+s} \right) ds$$

$$= r_2 + \frac{r_1 - r_2}{t} \log(1+t).$$

Consequently, the present value function is

$$P(t) = \exp\{-t\bar{r}(t)\}$$

$$= \exp\{-r_2 t\} \exp\{-\log((1+t)^{r_1-r_2})\}$$

$$= \exp\{-r_2 t\}(1+t)^{r_2-r_1}. \qquad \square$$

4.5 Exercises

Exercise 4.1 What is the effective interest rate when the nominal interest rate of 10% is

(a) compounded semiannually;
(b) compounded quarterly;
(c) compounded continuously?

Exercise 4.2 Suppose that you deposit your money in a bank that pays interest at a nominal rate of 10% per year. How long will it take for your money to double if the interest is compounded continuously?

Exercise 4.3 If you receive 5% interest compounded yearly, approximately how many years will it take for your money to quadruple? What if you were earning only 4%?

Exercise 4.4   Give a formula that approximates the number of years it would take for your funds to triple if you received interest at a rate r compounded yearly.

Exercise 4.5   How much do you need to invest at the beginning of each of the next 60 months in order to have a value of $100,000 at the end of 60 months, given that the annual nominal interest rate will be fixed at 6% and will be compounded monthly?

Exercise 4.6   The yearly cash flows of an investment are

$$-1,000, \ -1,200, \ 800, \ 900, \ 800.$$

Is this a worthwhile investment for someone who can both borrow and save money at the yearly interest rate of 6%?

Exercise 4.7   Consider two possible sequences of end-of-year returns:

$$20, \ 20, \ 20, \ 15, \ 10, \ 5 \quad \text{and} \quad 10, \ 10, \ 15, \ 20, \ 20, \ 20.$$

Which sequence is preferable if the interest rate, compounded annually, is: (a) 3%; (b) 5%; (c) 10%?

Exercise 4.8   A five-year $10,000 bond with a 10% coupon rate costs $10,000 and pays its holder $500 every six months for five years, with a final additional payment of $10,000 made at the end of those ten payments. Find its present value if the interest rate is: (a) 6%; (b) 10%; (c) 12%. Assume the compounding is monthly.

Exercise 4.9   A friend purchased a new sound system that was selling for $4,200. He agreed to make a down payment of $1,000 and to make 24 monthly payments of $160, beginning one month from the time of purchase. What is the effective interest rate being paid?

Exercise 4.10   Repeat Example 4.2b, this time assuming that the yearly interest rate is 20%.

Exercise 4.11   Repeat Example 4.2b, this time assuming that the cost of a new machine increases by $1,000 each year.

Exercise 4.12   Suppose you have agreed to a bank loan of $120,000, for which the bank charges no fees but 2 points. The quoted interest rate is .5% per month. You are required to pay only the accumulated interest each month for the next 36 months, at which point you must make a balloon payment of the still-owed $120,000. What is the effective interest rate of this loan?

Exercise 4.13 You can pay off a loan either by paying the entire amount of $16,000 now or you can pay $10,000 now and $10,000 at the end of ten years. Which is preferable when the nominal continuously compounded interest rate is: (a) 2%; (b) 5%; (c) 10%?

Exercise 4.14 A U.S. treasury bond (selling at a *par value* of $1,000) that matures at the end of five years is said to have a *coupon rate* of 6% if, after paying $1,000, the purchaser receives $30 at the end of each of the following nine six-month periods and then receives $1,030 at the end of the the tenth period. That is, the bond pays a simple interest rate of 3% per six-month period, with the principal repaid at the end of five years. Assuming a continuously compounded interest rate of 5%, find the present value of such a stream of cash payments.

Exercise 4.15 Explain why it is reasonable to suppose that $(1 + .05/n)^n$ is an increasing function of n for $n = 1, 2, 3, \ldots$.

Exercise 4.16 A bank pays a nominal interest rate of 6%, continuously compounded. If 100 is initially deposited, how much interest will be earned after

(a) 30 days;
(b) 60 days;
(c) 120 days?

Exercise 4.17 Assume continuously compounded interest at rate r. You plan to borrow 1,000 today, 2,000 one year from today, 3,000 two years from today, and then pay off all these loans three years from today. How much will you have to pay?

Exercise 4.18 The nominal interest rate is 5%, compounded yearly. How much would you have to pay today in order to receive the string of payments 3, 5, −6, 5, where the ith payment is to be received i years from now, $i = 1, 2, 3, 4$. (The payment −6 means that you will have to pay 6 three years from now.)

Exercise 4.19 Let r be the nominal interest rate, compounded yearly. For what values of r is the cash flow stream 20, 10 preferable to the cash flow stream 0, 34?

Exercise 4.20 What is the value of the continuously compounded nominal interest rate r if the present value of 104 to be received after 1 year is the same as the present value of 110 to be received after 2 years?

Exercise 4.21 Assuming continuously compounded interest at rate r, what is the present value of a cash flow sequence that returns the amount A at each of the times $s, s + t, s + 2t, \ldots$?

Exercise 4.22 Let $D(t)$ denote the amount you would have on deposit at time t if you deposit D at time 0 and interest is continuously compounded at rate r.

(a) Argue that, for h small, $D(t + h) \approx D(t) + rhD(t)$.
(b) Use (a) to argue that $D'(t) = rD(t)$.
(c) Use (b) to conclude that $D(t) = De^{rt}$.

Exercise 4.23 Consider two cash flow streams, where each will return the ith payment after i years:

$$100, 140, 131 \quad \text{and} \quad 90, 160, 120.$$

Is it possible to tell which cash flow stream is preferable without knowing the interest rate?

Exercise 4.24

(a) Find the yearly rate of return of an investment that, for an initial cost of 100, returns 110 after 2 years;
(b) Find the expected value of the yearly rate of return of an investment that, for an initial cost of 100, is equally likely to yield either 120 or 100 after 2 years.

Exercise 4.25 A zero coupon rate bond having face value F pays the bondholder the amount F when the bond matures. Assuming a continuously compounded interest rate of 8%, find the present value of a zero coupon bond with face value $F = 1,000$ that matures at the end of ten years.

Exercise 4.26 Find the rate of return for an investment that for an initial payment of 100 returns 40 at the end of 1 year and and additional 70 at the end of 2 years. What would the rate of return be if 70 were received after 1 year and 40 after 2 years?

Exercise 4.27

(a) Suppose for an initial investment of 1, you receive the nonnegative cash payments $x_1, \dots, x_n$, with x_i being received at the end of i periods. To determine if the rate of return of this investment is greater than 10 percent per period, is it necessary to first solve the equation $1 = \sum_{i=1}^{n} x_i (1+r)^{-i}$ for the rate of return r?

(b) For an initial investment of 100, an investor is to receive the amounts 8, 16, 110 at the end of the following three periods. Is the rate of return above 11 percent?

Exercise 4.28 For an initial investment of 100, an investment yields returns of X_i at the end of period i for $i = 1, 2$, where X_1 and X_2 are independent normal random variables with mean 60 and variance 25. What is the probability the rate of return of this investment is greater than 10 percent?

Exercise 4.29 The inflation rate is defined to be the rate at which prices as a whole are increasing. For instance, if the yearly inflation rate is 4% then what cost \$100 last year will cost \$104 this year. Let r_i denote the inflation rate, and consider an investment whose rate of return is r. We are often interested in determining the investment's rate of return from the point of view of how much the investment increases one's purchasing power; we call this quantity the investment's *inflation-adjusted rate of return* and denote it as r_a. Since the purchasing power of the amount $(1+r)x$ one year from now is equivalent to that of the amount $(1+r)x/(1+r_i)$ today, it follows that – with respect to constant purchasing power units – the investment transforms (in one time period) the amount x into the amount $(1+r)x/(1+r_i)$. Consequently, its inflation-adjusted rate of return is

$$r_a = \frac{1+r}{1+r_i} - 1.$$

When r and r_i are both small, we have the following approximation:

$$r_a \approx r - r_i.$$

For instance, if a bank pays a simple interest rate of 5% when the inflation rate is 3%, the inflation-adjusted interest rate is approximately 2%. What is its exact value?

Exercise 4.30 Consider an investment cash flow sequence $c_0, c_1, \ldots, c_n$, where $c_i < 0$, $i < n$, and $c_n > 0$. Show that if

$$P(r) = \sum_{i=0}^{n} c_i (1 + r)^{-i}$$

then, in the region $r > -1$,

(a) there is a unique solution of $P(r) = 0$;
(b) $P(r)$ need not be a monotone function of r.

Exercise 4.31 Suppose you can borrow money at an annual interest rate of 8% but can save money at an annual interest rate of only 5%. If you start with zero capital and if the yearly cash flows of an investment are

$$-1{,}000, \ 900, \ 800, \ -1{,}200, \ 700,$$

should you invest?

Exercise 4.32 Show that, if $r(t)$ is an nondecreasing function of t, then so is $\bar{r}(t)$.

Exercise 4.33 Show that the yield curve $\bar{r}(t)$ is a nondecreasing function of t if and only if

$$P(\alpha t) \geq (P(t))^{\alpha} \quad \text{for all } 0 \leq \alpha \leq 1, \ t \geq 0.$$

Exercise 4.34 Show that

$$\text{(a) } r(t) = -\frac{P'(t)}{P(t)} \quad \text{and} \quad \text{(b) } \bar{r}(t) = -\frac{\log P(t)}{t}.$$

Exercise 4.35 Plot the spot interest rate function $r(t)$ of Example 4.4a when

(a) $r_1 < r_2$;
(b) $r_2 < r_1$.

Reference Note: Proposition 4.2.1 is proven in Adler, Ilan and Sheldon M. Ross (2001). "A Probabilistic Approach to Identifying Positive Value Cash Flows," *The Mathematical Scientist*, 26.2

5. Pricing Contracts via Arbitrage

5.1   An Example in Options Pricing

Suppose that the nominal interest rate is r, and consider the following model for pricing an option to purchase a stock at a future time at a fixed price. Let the present price (in dollars) of the stock be 100 per share, and suppose we know that, after one time period, its price will be either 200 or 50 (see Figure 5.1). Suppose further that, for any y, at a cost of Cy you can purchase at time 0 the option to buy y shares of the stock at time 1 at a price of 150 per share. Thus, for instance, if you purchase this option and the stock rises to 200, then you would exercise the option at time 1 and realize a gain of $200 - 150 = 50$ for each of the y options purchased. On the other hand, if the price of the stock at time 1 is 50 then the option would be worthless. In addition to the options, you may also purchase x shares of the stock at time 0 at a cost of $100x$, and each share would be worth either 200 or 50 at time 1.

We will suppose that both x and y can be positive, negative, or zero. That is, you can either buy or sell both the stock and the option. For instance, if x were negative then you would be selling $-x$ shares of stock, yielding you an initial return of $-100x$, and you would then be responsible for buying and returning $-x$ shares of the stock at time 1 at a (time-1) cost of either 200 or 50 per share. (When you sell a stock that you do not own, we say that you are *selling it short*.)

We are interested in determining the appropriate value of C, the unit cost of an option. Specifically, we will show that if r is the one-period interest rate then, unless $C = [100 - 50(1 + r)^{-1}]/3$, there is a combination of purchases that will always result in a positive present value gain. To show this, suppose that at time 0 we

purchase x units of stock

and

purchase y units of options,

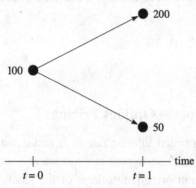

t = 0 \qquad\qquad t = 1$

Figure 5.1: Possible Stock Prices at Time 1

where x and y (both of which can be either positive or negative) are to be determined. The cost of this transaction is $100x + Cy$. If this amount is positive, then it should be borrowed from a bank, to be repaid with interest at time 1; if it is negative, then the amount received, $-(100x + Cy)$, should be put in the bank to be withdrawn at time 1. The value of our holdings at time 1 depends on the price of the stock at that time and is given by

$$\text{value} = \begin{cases} 200x + 50y & \text{if the price is 200,} \\ 50x & \text{if the price is 50.} \end{cases}$$

This formula follows by noting that, if the stock's price at time 1 is 200, then the x shares of the stock are worth $200x$ and the y units of options to buy the stock at a share price of 150 are worth $(200 - 150)y$. On the other hand, if the stock's price is 50, then the x shares are worth $50x$ and the y units of options are worthless. Now, suppose we choose y so that the value of our holdings at time 1 is the same no matter what the price of the stock at that time. That is, we choose y so that

$$200x + 50y = 50x$$

or

$$y = -3x.$$

Note that y has the opposite sign of x; thus, if $x > 0$ and so x shares of the stock are purchased at time 0, then $3x$ units of stock options are also *sold* at that time. Similarly, if x is negative, then $-x$ shares are sold and $-3x$ units of stock options are purchased at time 0.

Thus, with $y = -3x$, the

$$\text{time-1 value of holdings} = 50x$$

no matter what the value of the stock. As a result, if $y = -3x$ it follows that, after paying off our loan (if $100x + Cy > 0$) or withdrawing our money from the bank (if $100x + Cy < 0$), we will have gained the amount

$$\text{gain} = 50x - (100x + Cy)(1 + r)$$

$$= 50x - (100x - 3xC)(1 + r)$$

$$= (1 + r)x[3C - 100 + 50(1 + r)^{-1}].$$

Thus, if $3C = 100 - 50(1 + r)^{-1}$, then the gain is 0. On the other hand, if $3C \neq 100 - 50(1 + r)^{-1}$, then we can guarantee a positive gain (no matter what the price of the stock at time 1) by letting x be positive when $3C > 100 - 50(1 + r)^{-1}$ and by letting x be negative when $3C < 100 - 50(1 + r)^{-1}$.

For instance, if $(1 + r)^{-1} = .9$ and the cost per option is $C = 20$, then purchasing one share of the stock and selling three units of options initially costs us $100 - 3(20) = 40$, which is borrowed from the bank. However, the value of this holding at time 1 is 50 whether the stock price rises to 200 or falls to 50. Using $40(1 + r) = 44.44$ of this amount to pay our bank loan results in a guaranteed gain of 5.56. Similarly, if the cost of an option is 15, then selling one share of the stock ($x = -1$) and buying three units of options results in an initial gain of $100 - 45 = 55$, which is put into a bank to be worth $55(1 + r) = 61.11$ at time 1. Because the value of our holding at time 1 is -50, a guaranteed profit of 11.11 is attained. A sure-win betting scheme is called an *arbitrage*. Thus, for the numbers considered, the only option cost C that does not result in an arbitrage is $C = (100 - 45)/3 = 55/3$.

The existence of an arbitrage can often be seen by applying the law of one price.

Proposition 5.1.1 (The Law of One Price) *Consider two investments, the first of which costs the fixed amount C_1 and the second the fixed* amount C_2. *If the (present value) payoff from the first investment is always identical to that of the second investment, then either $C_1 = C_2$ or there is an arbitrage.*

The proof of the law of one price is immediate, because if their costs are unequal then an arbitrage is obtained by buying the cheaper investment and selling the more expensive one.

To apply the law of one price to our previous example, note that the payoff at time 1 from the investment of purchasing the call option is

$$\text{payoff of option} = \begin{cases} 50 & \text{if the price is 200,} \\ 0 & \text{if the price is 50.} \end{cases}$$

Consider now a second investment that calls for purchasing y shares of the security by borrowing x from the bank – to be repaid (with interest) at time 1 – and investing $100y - x$ of your own funds. Thus, the initial cost of this investment is $100y - x$. The payoff at time 1 from this investment is

$$\text{payoff of investment} = \begin{cases} 200y - x(1+r) & \text{if the price is 200,} \\ 50y - x(1+r) & \text{if the price is 50.} \end{cases}$$

Thus, if we choose x and y so that

$$200y - x(1+r) = 50,$$
$$50y - x(1+r) = 0,$$

then the payoffs from this investment and the option would be identical. Solving the preceding equations gives the solution

$$y = \frac{1}{3}, \quad x = \frac{50}{3(1+r)}.$$

Because the cost of the investment when using these values of x and y is $100y - x = \left(100 - \frac{50}{1+r}\right)/3$, it follows from the law of one price that either this is the cost of the option or there is an arbitage.

It is easy to specify the arbitrage (buy the cheaper investment and sell the more expensive one) when C, the cost of the option, is unequal to $\left(100 - \frac{50}{1+r}\right)/3$. Let us now do so.

Case 1: $C < \left(100 - \frac{50}{1+r}\right)/3$.

In this case sell $1/3$ share. Of the $100/3$ that this yields, use C to purchase an option and put the remainder (which is greater than $\frac{50}{3(1+r)}$) in the bank.

If the price at time 1 is 200, then your option will be worth 50 and you will have more than $50/3$ in the bank. Consequently you will have more than enough to meet your obligation of $200/3$ (which resulted from your short selling of $1/3$ share.) If the price at time 1 is 50 then you will have more than $50/3$ in the bank, which is more than enough to cover your obligation of $50/3$.

Case 2: $C > \left(100 - \frac{50}{1+r}\right)/3$.

In this case, sell the call, borrow $\frac{50}{3(1+r)}$ from the bank, and use $100/3$ of the amount received to purchase $1/3$ of a share. (The amount left over, $C - \left(100 - \frac{50}{1+r}\right)/3$, will be your arbitrage.) If the price at time 1 is 200, use the $200/3$ from your $1/3$ share to make the payments of $50/3$ to the bank and 50 to the call option buyer. If the price at time 1 is 50 then the option you sold is worthless, so use the $50/3$ from your $1/3$ share to pay the bank.

Remark. It should be noted that we have assumed, and will continue to do so unless otherwise noted, that there is always a market – in the sense that any investment can always be either bought or sold.

5.2 Other Examples of Pricing via Arbitrage

The type of option considered in Section 5.1 is known as a *call* option because it gives one the option of calling for the stock at a specified price, known as the *exercise* or *strike* price. An *American style* call option allows the buyer to exercise the option at any time up to the expiration time, whereas a *European style* call option can only be exercised at the expiration time. Although it might seem that, because of its additional flexibility, the American style option would be worth more, it turns out that it is never optimal to exercise a call option early; thus, the two style options have identical worths. We now prove this claim.

Proposition 5.2.1 *One should never exercise an American style call option before its expiration time t.*

Proof. Suppose that the present price of the stock is S, that you own an option to buy one share of the stock at a fixed price K, and that the option expires after an additional time t. If you exercise the option at this moment, you will realize the amount $S - K$. However, consider what would transpire if, instead of exercising the option, you sell the stock short and then purchase the stock at time t, either by paying the market price at that time or by exercising your option and paying K, whichever is less expensive. Under this strategy, you will initially receive S and will then have to pay the minimum of the market price and the exercise price K after an additional time t. This is clearly preferable to receiving S and immediately paying out K. □

In addition to call options there are also *put* options on stocks. These give their owners the option of putting a stock up for sale at a specified price. An American style put option allows the owner to put the stock up for sale – that is, to exercise the option – at any time up to the expiration time of the option. A European style put option can only be exercised at its expiration time. Contrary to the situation with call options, it may be advantageous to exercise a put option before its expiration time, and so the American style put option may be worth more than the European. The absence of arbitrage implies a relationship between the price of a European put option having exercise price K and expiration time t and the price of a call option on that stock that also has exercise price K and expiration time t. This is known as the *put–call option parity formula* and is as follows.

Proposition 5.2.2 *Let C be the price of a call option that enables its holder to buy one share of a stock at an exercise price K at time t; also, let P be the price of a European put option that enables its holder to sell one share of the stock for the amount K at time t. Let S be the price of the stock at time* 0. *Then, assuming that interest is continuously discounted at a nominal rate r, either*

$$S + P - C = Ke^{-rt}$$

or there is an arbitrage opportunity.

Proof. If

$$S + P - C < Ke^{-rt}$$

then we can effect a sure win by initially buying one share of the stock, buying one put option, and selling one call option. This initial payout of $S + P - C$ is borrowed from a bank to be repaid at time t. Let us now consider the value of our holdings at time t. There are two cases that depend on $S(t)$, the stock's market price at time t. If $S(t) \leq K$, then the call option we sold is worthless and we can exercise our put option to sell the stock for the amount K. On the other hand, if $S(t) > K$ then our put option is worthless and the call option we sold will be exercised, forcing us to sell our stock for the price K. Thus, in either case we will realize the amount K at time t. Since $K > e^{rt}(S + P - C)$, we can pay off our bank loan and realize a positive profit in all cases.

When

$$S + P - C > Ke^{-rt},$$

we can make a sure profit by reversing the procedure just described. Namely, we now sell one share of stock, sell one put option, and buy one call option. We leave the details of the verification to the reader. □

The arbitrage principle also determines the relationship between the present price of a stock and the contracted price to buy the stock at a specified time in the future. Our next two examples are related to these *forwards contracts*.

Example 5.2a *Forwards Contracts* Let S be the present market price of a specified stock. In a forwards agreement, one agrees at time 0 to pay the amount F at time t for one share of the stock that will be delivered at the time of payment. That is, one contracts a price for the stock, which is to be delivered and paid for at time t. We will now present an arbitrage argument to show that if interest is continuously discounted at the nominal interest rate r, then in order for there to be no arbitrage opportunity we must have

$$F = Se^{rt}.$$

To see why this equality must hold, suppose first that instead

$$F < Se^{rt}.$$

In this case, a sure win is obtained by selling the stock at time 0 with the understanding that you will buy it back at time t. Put the sale proceeds

S into a bond that matures at time t and, in addition, buy a forwards contract for delivery of one share of the stock at time t. Thus, at time t you will receive Se^{rt} from your bond. From this, you pay F to obtain one share of the stock, which you then return to settle your obligation. You thus end with a positive profit of $Se^{rt} - F$. On the other hand, if

$$F > Se^{rt}$$

then you can guarantee a profit of $F - Se^{rt}$ by simultaneously selling a forwards contract and borrowing S to purchase the stock. At time t you will receive F for your stock, out of which you repay your loan amount of Se^{rt}.   □

Remark. Another way to see that $F = Se^{rt}$ in the preceding example is to use the law of one price. Consider the following investments, both of which result in owning the security at time t.

(1) Put Fe^{-rt} in the bank and purchase a forward contract.
(2) Buy the security.

Thus, by the law of one price, either $Fe^{-rt} = S$ or there is an arbitrage.

When one purchases a share of a stock in the stock market, one is purchasing a share of ownership in the entity that issues the stock. On the other hand, the commodity market deals with more concrete objects: agricultural items like oats, corn, or wheat; energy products like crude oil and natural gas; metals such as gold, silver, or platinum; animal parts such as hogs, pork-bellies, and beef; and so on. Almost all of the activity on the commodities market is involved with contracts for future purchases and sales of the commodity. Thus, for instance, you could purchase a contract to buy natural gas in 90 days for a price that is specified today. (Such a *futures contract* differs from a forwards contract in that, although one pays in full when delivery is taken for both, in futures contracts one settles up on a daily basis depending on the change of the price of the futures contract on the commodity exchange.) You could also write a futures contract that obligates you to sell gas at a specified price at a specified time. Most people who play the commodities market never have actual contact with the commodity. Rather, people who buy a futures contract most often sell that contract before the delivery date. However, the relationship given in Example 5.2a does not hold for futures contracts in the commodity market. For one thing, if $F > Se^{rt}$ and you purchase the commodity (say, crude oil) to sell back at time t, then you will incur additional costs related to storing and insuring the oil. Also when $F < Se^{rt}$, to sell the commodity for today's price requires that you be able to deliver it immediately.

One of the most popular types of forward contracts involves currency exchanges, the topic of our next example.

Example 5.2b   The September 4, 1998, edition of the *New York Times* gives the following listing for the price of a German mark (or DM):

- today: .5777;
- 90-day forward: .5808.

In other words, you can purchase 1 DM today at the price of $.5777. In addition, you can sign a contract to purchase 1 DM in 90 days at a price, to be paid on delivery, of $.5808. Why are these prices different?

Solution.   One might suppose that the difference is caused by the market's expectation of the worth in 90 days of the German DM relative to the U.S. dollar, but it turns out that the entire price differential is due to the different interest rates in Germany and in the United States. Suppose that interest in both countries is continuously compounded at nominal yearly rates: r_u in the United States and r_g in Germany. Let S denote the present price of 1 DM, and let F be the price for a forwards contract to be delivered at time t. (This example considers the special case where $S = .5777$, $F = .5808$, and $t = 90/365$.) We now argue that, in order for there not to be an arbitrage opportunity, we must have

$$F = Se^{(r_u - r_g)t}.$$

To see why, consider two ways to obtain 1 DM at time t.

(1) Put $Fe^{-r_u t}$ in a U.S. bank and buy a forward contract to purchase 1 DM at time t.

(2) Purchase $e^{-r_g t}$ marks and put them in a German bank.

Note that the first investment, which costs $Fe^{-r_u t}$, and the second, which costs $Se^{-r_g t}$, both yield 1 DM at time t. Therefore, by the law of one price, either $Fe^{-r_u t} = Se^{-r_g t}$ or there is an arbitrage.

When $Fe^{-r_u t} < Se^{-r_g t}$, an arbitrage is obtained by borrowing 1 DM from a German bank, selling it for S U.S. dollars, and then putting that amount in a U.S. bank. At the same time, buy a forward contract to purchase $e^{r_g t}$ marks at time t. At time t, you will have $Se^{r_u t}$ dollars. Use $Fe^{r_g t}$ of this amount to pay the forward contract for $e^{r_g t}$ marks; then give these marks to the German bank to pay off your loan. Since $Se^{r_u t} > Fe^{r_g t}$, you have a positive amount remaining.

When $Fe^{-r_u t} > Se^{-r_g t}$, an arbitrage is obtained by borrowing $Se^{-r_g t}$ dollars from a U.S. bank and then using them to purchase $e^{-r_g t}$ marks, which are put in a German bank. Simultaneously, sell a forward contract for the purchase of 1 DM at time t. At time t, take out your 1 DM from the German bank and give it to the buyer of the forward contract, who will pay you F. Because $Se^{-r_g t}e^{r_u t}$ (the amount you must pay the U.S. bank to settle your loan) is less than F, you have an arbitrage. □

The following is an obvious generalization of the law of one price.

Proposition 5.2.3 (The Generalized Law of One Price) *Consider two investments, the first of which costs the fixed amount C_1 and the second the fixed amount C_2. If $C_1 < C_2$ and the (present value) payoff from the first investment is always at least as large as that from the second investment, then there is an abitrage.*

The arbitrage is clearly obtained by simultaneously buying investment 1 and selling investment 2.

Before applying the generalized law of one price, we need the following definition.

Definition A function $f(x)$ is said to be *convex* if, for for all x and y and $0 < \lambda < 1$,

$$f(\lambda x + (1 - \lambda)y) \leq \lambda f(x) + (1 - \lambda)f(y).$$

For a geometric interpretation of convexity, note that $\lambda f(x) + (1-\lambda)f(y)$ is a point on the straight line between $f(x)$ and $f(y)$ that is as much weighted toward $f(x)$ as is the point $\lambda x + (1 - \lambda)y$ on the straight line between x and y weighted toward x. Consequently, convexity can be interpreted as stating that the straight line segment connecting two points on the curve $f(x)$ always lies above (or on) the curve (Figure 5.2).

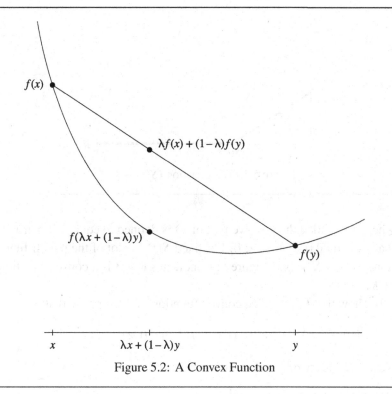

$f(x)$

$\lambda f(x) + (1-\lambda)f(y)$

$f(\lambda x + (1-\lambda)y)$

$f(y)$

x $\lambda x + (1-\lambda)y$ y

Figure 5.2: A Convex Function

Proposition 5.2.4 *Let $C(K, t)$ be the cost of a call option on a specified security that has strike price K and expiration time t.*

(a) *For fixed expiration time t, $C(K, t)$ is a convex and nonincreasing function of K.*

(b) *For $s > 0$, $C(K, t) - C(K + s, t) \leq se^{-rt}$.*

Proof. If $S(t)$ denotes the price of the security at time t, then the payoff at time t from a (K, t) call option is

$$\text{payoff of option} = \begin{cases} S(t) - K & \text{if } S(t) \geq K, \\ 0 & \text{if } S(t) < K. \end{cases}$$

That is,

$$\text{payoff of option} = (S(t) - K)^{+},$$

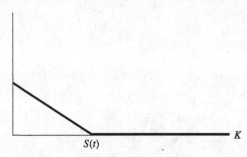

Figure 5.3: The Function $(S(t) - K)^+$

where x^+ (called the positive part of x) is defined to equal x when $x \geq 0$ and to equal 0 when $x < 0$. For fixed $S(t)$, a plot of the payoff function $(S(t) - K)^+$ (see Figure 5.3) indicates that it is a convex function of K.

To show that $C(K, t)$ is a convex function of K, suppose that

$$K = \lambda K_1 + (1 - \lambda)K_2 \quad \text{for } 0 < \lambda < 1.$$

Now consider two investments:

(1) purchase a (K, t) call option;
(2) purchase λ (K_1, t) call options and $1 - \lambda$ (K_2, t) call options.

Because the payoff at time t from investment (1) is $(S(t) - K)^+$ whereas that from investment (2) is $\lambda(S(t) - K_1)^+ + (1 - \lambda)(S(t) - K_2)^+$, it follows from the convexity of the function $(S(t) - K)^+$ that the payoff from investment (2) is at least as large as that from investment (1). Consequently, by the generalized law of one price, either the cost of investment (2) is at least as large as that of investment (1) or there is an arbitrage. That is, either

$$C(K, t) \leq \lambda C(K_1, t) + (1 - \lambda)C(K_2, t)$$

or there is an arbitrage. Hence, convexity is established. The proof that $C(K, t)$ is nonincreasing in K is left as an exercise.

To prove part (b), note that if $C(K, t) > C(K + s, t) + se^{-rt}$ then an arbitrage is possible by selling a call with strike price K and exercise time t, buy a $(K + s, t)$ call, and put the remaining amount

$C(K, t) - C(K + s, t) \geq se^{-rt}$ in the Bank. Because the payoff of the call with strike price K can exceed that of the one with price $K + s$ by at most s, this combination of buying one call and selling the other always yields a positive profit. □

Remark. Part (b) of Proposition 5.2.4 is equivalent to the statement that

$$\frac{\partial}{\partial K} C(K, t) \geq -e^{-rt}. \tag{5.1}$$

To see why they are equivalent, note that (b) implies

$$C(K + s, t) - C(K, t) \geq -se^{-rt} \quad \text{for } s > 0.$$

Dividing both sides of this inequality by s and letting s go to 0 then yields the result. To show that the inequality (5.1) implies Proposition 5.2.4(b), suppose (5.1) holds. Then

$$\int_K^{K+s} \frac{\partial}{\partial x} C(x, t)\, dx \geq \int_K^{K+s} -e^{-rt} dx,$$

showing that

$$C(K + s, t) - C(K, t) \geq -se^{-rt},$$

which is part (b).

Our next example uses the generalized law of one price to show that an option on an index – defined as a weighted sum of the prices of a collection of specified securities – will never be more expensive than the costs of a corresponding collection of options on the individual securities. This result is sometimes called the *option portfolio property*.

Example 5.2c Consider a collection of n securities, and for $j = 1, \ldots, n$ let $S_j(y)$ denote the price of security j at a time y in the future. For fixed positive constants w_j, let

$$I(y) = \sum_{j=1}^n w_j S_j(y).$$

That is, $I(y)$ is the market value at time y of a portfolio of the securities, where the portfolio consists of w_j shares of security j. Let a (K_j, t) call option on security j refer to a call option having strike price K_j and expiration time t, and let C_j ($j = 1, \ldots, n$) denote the costs of these options. Also, let C be the cost of a call option on the index I that has strike price $\sum_{j=1}^{n} w_j K_j$ and expiration time t. We now show that the payoff of the call option on the index is always less than or equal to the sum of the payoffs from buying w_j (K_j, t) call options on security j for each $j = 1, \ldots, n$:

index option payoff at time t

$$= \left(I(t) - \sum_{j=1}^{n} w_j K_j \right)^{+}$$

$$= \left(\sum_{j=1}^{n} w_j S_j(t) - \sum_{j=1}^{n} w_j K_j \right)^{+}$$

$$= \left(\sum_{j=1}^{n} w_j (S_j(t) - K_j) \right)^{+}$$

$$\leq \left(\sum_{j=1}^{n} (w_j (S_j(t) - K_j))^{+} \right)^{+} \qquad \text{(because } x \leq x^{+})$$

$$= \left(\sum_{j=1}^{n} w_j (S_j(t) - K_j)^{+} \right)^{+}$$

$$= \sum_{j=1}^{n} w_j (S_j(t) - K_j)^{+}$$

$$= \sum_{j=1}^{n} w_j \cdot [\text{payoff from } (K_j, t) \text{ call option}].$$

Consequently, by the generalized law of one price, we have that either $C \leq \sum_{j=1}^{n} w_j C_j$ or there is an arbitrage.   □

5.3   Exercises

Exercise 5.1   Suppose you pay 10 to buy a European $(K = 100, t = 2)$ call option on a given security. Assuming a continuously compounded nominal annual interest rate of 6 percent, find the present value of your return from this investment if the price of the security at time 2 is

(a) 110;

(b) 98.

Exercise 5.2 Suppose you pay 5 to buy a European $(K = 100,$ $t = 1/2)$ put option on a given security. Assuming a nominal annual interest rate of 6 percent, compounded monthly, find the present value of your return from this investment if

(a) $S(1/2) = 102$;
(b) $S(1/2) = 98$.

Exercise 5.3 Suppose it is known that the price of a certain security after one period will be one of the m values $s_1, \ldots, s_m$. What should be the cost of an option to purchase the security at time 1 for the price K when $K < \min s_i$?

Exercise 5.4 Let C be the price of a call option to purchase a security whose present price is S. Argue that $C \leq S$.

Exercise 5.5 Let C be the cost of a call option to purchase a security at time t for the price K. Let S be the current price of the security, and let r be the interest rate. State and prove an inequality involving the quantities C, S, and Ke^{-rt}.

Exercise 5.6 The current price of a security is 30. Given an interest rate of 5%, compounded continuously, find a lower bound for the price of a call option that expires in four months and has a strike price of 28.

Exercise 5.7 Let P be the price of a put option to sell a security, whose present price is S, for the amount K. Which of the following are necessarily true?

(a) $P \leq S$.
(b) $P \leq K$.

Exercise 5.8 Let P be the price of a put option to sell a security, whose present price is S, for the amount K. Argue that

$$P \geq Ke^{-rt} - S,$$

where t is the exercise time and r is the interest rate.

Exercise 5.9 With regard to Proposition 5.2.2, verify that the strategy of selling one share of stock, selling one put option, and buying one call option always results in a positive win if $S + P - C > Ke^{-rt}$.

Exercise 5.10   Use the law of one price to prove the put–call option parity formula.

Exercise 5.11   The current price of a security is s. Suppose that its possible prices at time t are s_1 or s_2. Consider a K, t European put option on this security, and suppose that $K > s_1 > s_2$.

(a) If you buy the put and the security, what is your return at time t?
(b) What is the no-arbitrage cost of the put?

Exercise 5.12   A digital (K, t) call option gives its holder 1 at expiration time t if $S(t) \geq K$, or 0 if $S(t) < K$. A digital (K, t) put option gives its holder 1 at expiration time t if $S(t) < k$, or 0 if $S(t) \geq K$. Let C_1 and C_2 be the costs of such digital call and put options on the same security. Derive a put-call parity relationship between C_1 and C_2.

Exercise 5.13   A European call and put option on the same security both expire in three months, both have a strike price of 20, and both sell for the price 3. If the nominal continuously compounded interest rate is 10% and the stock price is currently 25, identify an arbitrage.

Exercise 5.14   Let C_a and P_a be the costs of American call and put options (respectively) on the same security, both having the same strike price K and exercise time t. If S is the present price of the security, give either an identity or an inequality that relates the quantities C_a, P_a, K, and e^{-rt}. Briefly explain.

Exercise 5.15   Consider two put options on the same security, both of which have expiration t. Suppose the exercise prices of the two puts are K_1 and K_2, where $K_1 > K_2$. Argue that

$$K_1 - K_2 \geq P_1 - P_2,$$

where P_i is the price of the put with strike K_i, $i = 1, 2$.

Exercise 5.16   Explain why the price of an American put option having exercise time t cannot be less than the price of a second put option on the same security that is identical to the first option except that its exercise time is earlier.

Exercise 5.17 Say whether each of the following statements is always true, always false, or sometimes true and sometimes false. Assume that, aside from what is mentioned, all other parameters remain fixed. Give brief explanations for your answers.

(a) The price of a European call option is nondecreasing in its expiration time.
(b) The price of a forward contract on a foreign currency is nondecreasing in its maturity date.
(c) The price of a European put option is nondecreasing in its expiration time.

Exercise 5.18 Your financial adviser has suggested that you buy both a European put and a European call on the same security, with both options expiring in three months, and both having a strike price equal to the present price of the security.

(a) Under what conditions would such an investment strategy seem reasonable?
(b) Plot the return at time $t = 1/4$ from this strategy as a function of the price of the security at that time.

Exercise 5.19 If a stock is selling for a price s immediately before it pays a dividend d (i.e., the amount d per share is paid to every shareholder), then what should its price be immediately after the dividend is paid?

Exercise 5.20 Let $S(t)$ be the price of a given security at time t. All of the following options have exercise time t and, unless stated otherwise, exercise price K. Give the payoff at time t that is earned by an investor who:

(a) owns one call and one put option;
(b) owns one call having exercise price K_1 and has sold one put having exercise price K_2;
(c) owns two calls and has sold short one share of the security;
(d) owns one share of the security and has sold one call.

Exercise 5.21 Argue that the price of a European call option is nonincreasing in its strike price.

Exercise 5.22 Suppose that you simultaneously buy a call option with strike price 100 and write (i.e., sell) a call option with strike price 105 on the same security, with both options having the same expiration time.

(a) Is your initial cost positive or negative?
(b) Plot your return at expiration time as a function of the price of the security at that time.

Exercise 5.23 Consider two call options on a security whose present price is 110. Suppose that both call options have the same expiration time; one has strike price 100 and costs 20, whereas the other has strike price 110 and costs C. Assuming that an arbitrage is not possible, give a lower bound on C.

Exercise 5.24 Let $P(K, t)$ denote the cost of a European put option with strike K and expiration time t. Prove that $P(K, t)$ is convex in K for fixed t, or explain why it is not necessarily true.

Exercise 5.25 Can the proof given in the text for the cost of a call option be modified to show that the cost of an American put option is convex in its strike price?

Exercise 5.26 A (K_1, t_1, K_2, t_2) *double call option* is one that can be exercised either at time t_1 with strike price K_1 or at time t_2 $(t_2 > t_1)$ with strike price K_2. Argue that you would never exercise at time t_1 if $K_1 > e^{-r(t_2-t_1)} K_2$.

Exercise 5.27 In a *capped call option*, the return is capped at a certain specified value A. That is, if the option has strike price K and expiration time t, then the payoff at time t is

$$\min(A, (S(t) - K)^+),$$

where $S(t)$ is the price of the security at time t. Show that an equivalent way of defining such an option is to let

$$\max(K, S(t) - A)$$

be the strike price when the call is exercised at time t.

Exercise 5.28 Argue that an American capped call option should be exercised early only when the price of the security is at least $K + A$.

Exercise 5.29 A function $f(x)$ is said to be *concave* if, for all x, y and $0 < \lambda < 1$,

$$f(\lambda x + (1 - \lambda)y) \geq \lambda f(x) + (1 - \lambda)f(y).$$

(a) Give a geometrical interpretation of when a function is concave.
(b) Argue that $f(x)$ is concave if and only if $g(x) = -f(x)$ is convex.

Exercise 5.30 Consider two investments, where investment i, $i = 1, 2$, costs C_i and yields the return X_i after 1 year, where X_1 and X_2 are random variables. Suppose $C_1 > C_2$. Are the following statements necessarily true?

(a) If $E[X_1] < E[X_2]$, then there is an arbitrage.
(b) If $P\{X_2 > X_1\} > 0$, then there is an arbitrage.

REFERENCES

[1] Cox, J., and M. Rubinstein (1985). *Options Markets.* Englewood Cliffs, NJ: Prentice-Hall.
[2] Merton, R. (1973). "Theory of Rational Option Pricing." *Bell Journal of Economics and Mangagement Science* 4: 141–83.
[3] Samuelson, P., and R. Merton (1969). "A Complete Model of Warrant Pricing that Maximizes Utility." *Industrial Management Review* 10: 17–46.
[4] Stoll, H. R., and R. E. Whaley (1986). "New Option Intruments: Arbitrageable Linkages and Valuation." *Advances in Futures and Options Research* 1 (part A): 25–62.

6. The Arbitrage Theorem

6.1 The Arbitrage Theorem

Consider an experiment whose set of possible outcomes is $\{1, 2, \ldots, m\}$, and suppose that n wagers concerning this experiment are available. If the amount x is bet on wager i, then $x r_i(j)$ is received if the outcome of the experiment is j ($j = 1, \ldots, m$). In other words, $r_i(\cdot)$ is the *return function* for a unit bet on wager i. The amount bet on a wager is allowed to be positive, negative, or zero.

A betting strategy is a vector $\mathbf{x} = (x_1, x_2, \ldots, x_n)$, with the interpretation that x_1 is bet on wager 1, x_2 is bet on wager 2, $\ldots$, x_n is bet on wager n. If the outcome of the experiment is j, then the return from the betting strategy $\mathbf{x}$ is given by

$$\text{return from } \mathbf{x} = \sum_{i=1}^{n} x_i r_i(j).$$

The following result, known as the *arbitrage theorem,* states that either there exists a probability vector $\mathbf{p} = (p_1, p_2, \ldots, p_m)$ on the set of possible outcomes of the experiment under which the expected return of each wager is equal to zero, or else there exists a betting strategy that yields a positive win for each outcome of the experiment.

Theorem 6.1.1 (The Arbitrage Theorem) *Exactly one of the following is true: Either*

(a) *there is a probability vector* $\mathbf{p} = (p_1, p_2, \ldots, p_m)$ *for which*

$$\sum_{j=1}^{m} p_j r_i(j) = 0 \quad \textit{for all } i = 1, \ldots, n,$$

or else

(b) *there is a betting strategy* $\mathbf{x} = (x_1, x_2, \ldots, x_n)$ *for which*

$$\sum_{i=1}^{n} x_i r_i(j) > 0 \quad \text{for all } j = 1, \ldots, m.$$

Proof. See Section 6.3.

If X is the outcome of the experiment, then the arbitrage theorem states that either there is a set of probabilities $(p_1, p_2, \ldots, p_m)$ such that if

$$P\{X = j\} = p_j \quad \text{for all } j = 1, \ldots, m$$

then

$$E[r_i(X)] = 0 \quad \text{for all } i = 1, \ldots, n,$$

or else there is a betting strategy that leads to a sure win. In other words, either there is a probability vector on the outcomes of the experiment that results in all bets being fair, or else there is a betting scheme that guarantees a win.

Definition Probabilities on the set of outcomes of the experiment that result in all bets being fair are called *risk-neutral probabilities*.

Example 6.1a In some situations, the only type of wagers allowed are ones that choose one of the outcomes i ($i = 1, \ldots, m$) and then bet that i is the outcome of the experiment. The return from such a bet is often quoted in terms of *odds*. If the odds against outcome i are o_i (often expressed as "o_i to 1"), then a one-unit bet will return either o_i if i is the outcome of the experiment or -1 if i is not the outcome. That is, a one-unit bet on i will either win o_i or lose 1. The return function for such a bet is given by

$$r_i(j) = \begin{cases} o_i & \text{if } j = i, \\ -1 & \text{if } j \neq i. \end{cases}$$

Suppose that the odds $o_1, o_2, \ldots, o_m$ are quoted. In order for there not to be a sure win, there must be a probability vector $\mathbf{p} = (p_1, p_2, \ldots, p_m)$ such that, for each i ($i = 1, \ldots, m$),

$$0 = E_{\mathbf{p}}[r_i(X)] = o_i p_i - (1 - p_i).$$

That is, we must have

$$p_i = \frac{1}{1 + o_i}.$$

Since the p_i must sum to 1, this means that the condition for there not to be an arbitrage is that

$$\sum_{i=1}^{m} \frac{1}{1 + o_i} = 1.$$

That is, if $\sum_{i=1}^{m} (1 + o_i)^{-1} \neq 1$, then a sure win is possible. For instance, suppose there are three possible outcomes and the quoted odds are as follows.

Outcome	Odds
1	1
2	2
3	3

That is, the odds against outcome 1 are 1 to 1; they are 2 to 1 against outcome 2; and they are 3 to 1 against outcome 3. Since

$$\frac{1}{2} + \frac{1}{3} + \frac{1}{4} = \frac{13}{12} \neq 1,$$

a sure win is possible. One possibility is to bet -1 on outcome 1 (so you either win 1 if the outcome is not 1 or you lose 1 if the outcome is 1) and bet $-.7$ on outcome 2 (so you either win .7 if the outcome is not 2 or you lose 1.4 if it is 2), and $-.5$ on outcome 3 (so you either win .5 if the outcome is not 3 or you lose 1.5 if it is 3). If the experiment results in outcome 1, you win $-1 + .7 + .5 = .2$; if it results in outcome 2, you win $1 - 1.4 + .5 = .1$; if it results in outcome 3, you win $1 + .7 - 1.5 = .2$. Hence, in all cases you win a positive amount.   □

Example 6.1b Let us reconsider the option pricing example of Section 5.1, where the initial price of a stock is 100 and the price after one period is assumed to be either 200 or 50. At a cost of C per share, we can purchase at time 0 the option to buy the stock at time 1 for the price of 150. For what value of C is no sure win possible?

Solution. In the context of this section, the outcome of the experiment is the value of the stock at time 1; thus, there are two possible outcomes. There are also two different wagers: to buy (or sell) the stock, and to buy (or sell) the option. By the arbitrage theorem, there will be no sure win if there are probabilities $(p, 1 - p)$ on the outcomes that make the expected present value return equal to zero for both wagers.

The present value return from purchasing one share of the stock is

$$\text{return} = \begin{cases} 200(1 + r)^{-1} - 100 & \text{if the price is 200 at time 1,} \\ 50(1 + r)^{-1} - 100 & \text{if the price is 50 at time 1.} \end{cases}$$

Hence, if p is the probability that the price is 200 at time 1, then

$$E[\text{return}] = p\left[\frac{200}{1 + r} - 100\right] + (1 - p)\left[\frac{50}{1 + r} - 100\right]$$

$$= p\frac{150}{1 + r} + \frac{50}{1 + r} - 100.$$

Setting this equal to zero yields that

$$p = \frac{1 + 2r}{3}.$$

Therefore, the only probability vector $(p, 1 - p)$ that results in a zero expected return for the wager of purchasing the stock has $p = (1 + 2r)/3$.

In addition, the present value return from purchasing one option is

$$\text{return} = \begin{cases} 50(1 + r)^{-1} - C & \text{if the price is 200 at time 1,} \\ -C & \text{if the price is 50 at time 1.} \end{cases}$$

Hence, when $p = (1 + 2r)/3$, the expected return of purchasing one option is

$$E[\text{return}] = \frac{1 + 2r}{3}\frac{50}{1 + r} - C.$$

It thus follows from the arbitrage theorem that the only value of C for which there will not be a sure win is

$$C = \frac{1 + 2r}{3}\frac{50}{1 + r};$$

that is, when

$$C = \frac{50 + 100r}{3(1 + r)},$$

which is in accord with the result of Section 5.1.   □

6.2   The Multiperiod Binomial Model

Let us now consider a stock option scenario in which there are n periods and where the nominal interest rate is r per period. Let $S(0)$ be the initial price of the stock, and for $i = 1, \ldots, n$ let $S(i)$ be its price at i time periods later. Suppose that $S(i)$ is either $uS(i-1)$ or $dS(i-1)$, where $d < 1 + r < u$. That is, going from one time period to the next, the price either goes up by the factor u or down by the factor d. Furthermore, suppose that at time 0 an option may be purchased that enables one to buy the stock after n periods have passed for the amount K. In addition, the stock may be purchased and sold anytime within these n time periods.

Let X_i equal 1 if the stock's price goes up by the factor u from period $i - 1$ to i, and let it equal 0 if that price goes down by the factor d. That is,

$$X_i = \begin{cases} 1 & \text{if } S(i) = uS(i-1), \\ 0 & \text{if } S(i) = dS(i-1). \end{cases}$$

The outcome of the experiment can now be regarded as the value of the vector $(X_1, X_2, \ldots, X_n)$. It follows from the arbitrage theorem that, in order for there not to be an arbitrage opportunity, there must be probabilities on these outcomes that make all bets fair. That is, there must be a set of probabilities

$$P\{X_1 = x_1, \ldots, X_n = x_n\}, \quad x_i = 0, 1, \ i = 1, \ldots, n,$$

that make all bets fair.

Now consider the following type of bet: First choose a value of i ($i = 1, \ldots, n$) and a vector $(x_1, \ldots, x_{i-1})$ of zeros and ones, and then observe the first $i - 1$ changes. If $X_j = x_j$ for each $j = 1, \ldots, i - 1$, immediately buy one unit of stock and then sell it back the next period. If the stock is purchased, then its cost at time $i - 1$ is $S(i-1)$; the time-$(i-1)$ value of the amount obtained when it is then sold at time i is either $(1 + r)^{-1}uS(i-1)$ if the stock goes up or $(1 + r)^{-1}dS(i-1)$ if it goes down. Therefore, if we let

$$\alpha = P\{X_1 = x_1, \ldots, X_{i-1} = x_{i-1}\}$$

denote the probability that the stock is purchased, and let

$$p = P\{X_i = 1 \mid X_1 = x_1, \ldots, X_{i-1} = x_{i-1}\}$$

denote the probability that a purchased stock goes up the next period, then the expected gain on this bet (in time-$(i - 1)$ units) is

$$\alpha[p(1 + r)^{-1}uS(i - 1) + (1 - p)(1 + r)^{-1}dS(i - 1) - S(i - 1)].$$

Consequently, the expected gain on this bet will be zero, provided that

$$\frac{pu}{1 + r} + \frac{(1 - p)d}{1 + r} = 1$$

or, equivalently, that

$$p = \frac{1 + r - d}{u - d}.$$

In other words, the only probability vector that results in an expected gain of zero for this type of bet has

$$P\{X_i = 1 \mid X_1 = x_1, \ldots, X_{i-1} = x_{i-1}\} = \frac{1 + r - d}{u - d}.$$

Since $x_1, \ldots, x_n$ are arbitrary, this implies that the only probability vector on the set of outcomes that results in all these bets being fair is the one that takes $X_1, \ldots, X_n$ to be independent random variables with

$$P\{X_i = 1\} = p = 1 - P\{X_i = 0\}, \quad i = 1, \ldots, n, \qquad (6.1)$$

where

$$p = \frac{1 + r - d}{u - d}. \qquad (6.2)$$

It can be shown that, with these probabilities, any bet on buying stock will have zero expected gain. Thus, it follows from the arbitrage theorem that either the cost of the option must be equal to the expectation of the present (i.e., the time-0) value of owning it using the preceding probabilities, or else there will be an arbitrage opportunity. So, to determine the no-arbitrage cost, assume that the X_i are independent 0-or-1 random variables whose common probability p of being equal to 1 is given by Equation (6.2). Letting Y denote their sum, it follows that Y is just the number of the X_i that are equal to 1, and thus Y is a binomial random variable with parameters n and p. Now, in going from period to period, the stock's price is its old price multiplied either by u or by d. At time n, the price would have gone up Y times and down $n - Y$

times, so it follows that the stock's price after n periods can be expressed as

$$S(n) = u^Y d^{n-Y} S(0),$$

where $Y = \sum_{i=1}^{n} X_i$ is, as previously noted, a binomial random variable with parameters n and p. The value of owning the option after n periods have elapsed is $(S_{(n)} - K)^+$, which is defined to equal either $S_{(n)} - K$ (when this quantity is nonnegative) or zero (when it is negative). Therefore, the present (time-0) value of owning the option is

$$(1+r)^{-n}(S(n) - K)^+$$

and so the expectation of the present value of owning the option is

$$(1+r)^{-n} E[(S(n) - K)^+] = (1+r)^{-n} E[(S(0)u^Y d^{n-Y} - K)^+].$$

Thus, the only option cost C that does not result in an arbitrage is

$$C = (1+r)^{-n} E[(S(0)u^Y d^{n-Y} - K)^+]. \tag{6.3}$$

Remark. Although Equation (6.3) could be streamlined for computational convenience, the expression as given is sufficient for our main purpose: determining the unique no-arbitrage option cost when the underlying security follows a geometric Brownian motion. This is accomplished in our next chapter, where we derive the famous Black–Scholes formula.

6.3   Proof of the Arbitrage Theorem

In order to prove the arbitrage theorem, we first present the duality theorem of linear programming as follows. Suppose that, for given constants c_i, b_j, and $a_{i,j}$ ($i = 1, \ldots, n$, $j = 1, \ldots, m$), we want to choose values $x_1, \ldots, x_n$ that will

$$\text{maximize} \sum_{i=1}^{n} c_i x_i$$

$$subject\ to$$

$$\sum_{i=1}^{n} a_{i,j} x_i \le b_j, \quad j = 1, 2, \ldots, m.$$

This problem is called a primal linear program. Every primal linear program has a *dual* problem, and the dual of the preceding linear program is to choose values $y_1, \dots, y_m$ that

$$\text{minimize} \sum_{j=1}^{m} b_j y_j$$

subject to

$$\sum_{j=1}^{m} a_{i,j} y_j = c_i, \quad i = 1, \dots, n,$$

$$y_j \geq 0, \quad j = 1, \dots, m.$$

A linear program is said to be *feasible* if there are variables $(x_1, \dots, x_n$ in the primal linear program or $y_1, \dots, y_m$ in the dual) that satisfy the constraints. The key theoretical result of linear programming is the *duality theorem*, which we state without proof.

Proposition 6.3.1 (Duality Theorem of Linear Programming) *If a primal and its dual linear program are both feasible, then they both have optimal solutions and the maximal value of the primal is equal to the minimal value of the dual. If either problem is infeasible, then the other does not have an optimal solution.*

A consequence of the duality theorem is the arbitrage theorem. Recall that the arbitrage theorem refers to a situation in which there are n wagers with payoffs that are determined by the result of an experiment having possible outcomes $1, 2, \dots, m$. Specifically, if you bet wager i at level x, then you win the amount $x r_i(j)$ if the outcome of the experiment is j. A betting strategy is a vector $\mathbf{x} = (x_1, \dots, x_n)$, where each x_i can be positive or negative (or zero), and with the interpretation that you simultaneously bet wager i at level x_i for each $i = 1, \dots, n$. If the outcome of the experiment is j, then your winnings from the betting strategy $\mathbf{x}$ are

$$\sum_{i=1}^{n} x_i r_i(j).$$

Proposition 6.3.2 (Arbitrage Theorem) *Exactly one of the following is true: Either*

(i) *there exists a probability vector* $\mathbf{p} = (p_1, \ldots, p_m)$ *for which*

$$\sum_{j=1}^{m} p_j r_i(j) = 0 \quad \text{for all } i = 1, \ldots, n;$$

or

(ii) *there exists a betting strategy* $\mathbf{x} = (x_1, \ldots, x_n)$ *such that*

$$\sum_{i=1}^{n} x_i r_i(j) > 0 \quad \text{for all } j = 1, \ldots, m.$$

That is, either there exists a probability vector under which all wagers have expected gain equal to zero, or else there is a betting strategy that always results in a positive win.

Proof. Let x_{n+1} denote an amount that the gambler can be sure of winning, and consider the problem of maximizing this amount. If the gambler uses the betting strategy $(x_1, \ldots, x_n)$ then she will win $\sum_{i=1}^{n} x_i r_i(j)$ if the outcome of the experiment is j. Hence, she will want to choose her betting strategy $(x_1, \ldots, x_n)$ and x_{n+1} so as to

$$\text{maximize } x_{n+1}$$

subject to

$$\sum_{i=1}^{n} x_i r_i(j) \geq x_{n+1}, \quad j = 1, \ldots, m.$$

Letting

$$a_{i,j} = -r_i(j), \quad i = 1, \ldots, n, \quad a_{n+1,j} = 1,$$

we can rewrite the preceding as follows:

$$\text{maximize } x_{n+1}$$

subject to

$$\sum_{i=1}^{n+1} a_{i,j} x_i \leq 0, \quad j = 1, \ldots, m.$$

Note that the preceding linear program has $c_1 = c_2 = \cdots = c_n = 0$, $c_{n+1} = 1$, and upper-bound constraint values all equal to zero (i.e., all

$b_j = 0$). Consequently, its dual program is to choose variables $y_1, \ldots, y_m$ so as to

$$\text{minimize } 0$$

$$\textit{subject to}$$

$$\sum_{j=1}^{m} a_{i,j} y_j = 0, \quad i = 1, \ldots, n,$$

$$\sum_{j=1}^{m} a_{n+1,j} y_j = 1,$$

$$y_j \geq 0, \quad j = 1, \ldots, m.$$

Using the definitions of the quantities $a_{i,j}$ gives that this dual linear program can be written as

$$\text{minimize } 0$$

$$\textit{subject to}$$

$$\sum_{j=1}^{m} r_i(j) y_j = 0, \quad i = 1, \ldots, n,$$

$$\sum_{j=1}^{m} y_j = 1,$$

$$y_j \geq 0, \quad j = 1, \ldots, m.$$

Observe that this dual will be feasible, and its minimal value will be zero, if and only if there is a probability vector $(y_1, \ldots, y_m)$ under which all wagers have expected return 0. The primal problem is feasible because $x_i = 0$ $(i = 1, \ldots, n + 1)$ satisfies its constraints, so it follows from the duality theorem that if the dual problem is also feasible then the optimal value of the primal is zero and hence no sure win is possible. On the other hand, if the dual is infeasible then it follows from the duality theorem that there is no optimal solution of the primal. But this implies that zero is not the optimal solution, and thus there is a betting scheme whose minimal return is positive. (The reason there is no primal optimal solution when the dual is infeasible is because the primal is unbounded in this case. That is, if there is a betting scheme $\mathbf{x}$ that gives a guaranteed return of at least $v > 0$, then $c\mathbf{x}$ gives a guaranteed return of at least cv.) $\qquad \square$

6.4   Exercises

Exercise 6.1 Consider an experiment with three possible outcomes and odds as follows.

Outcome	Odds
1	1
2	2
3	5

Is there a betting scheme that results in a sure win?

Exercise 6.2 Consider an experiment with four possible outcomes, and suppose that the quoted odds for the first three of these outcomes are as follows.

Outcome	Odds
1	2
2	3
3	4

What must be the odds against outcome 4 if there is to be no possible arbitrage when one is allowed to bet both for and against any of the outcomes?

Exercise 6.3 An experiment can result in any of the outcomes 1, 2, or 3.

(a) If there are two different wagers, with

$$r_1(1) = 4, \ r_1(2) = 8, \ r_1(3) = -10$$

$$r_2(1) = 6, \ r_2(2) = 12, \ r_2(3) = -16$$

is an arbitrage possible?

(b) If there are three different wagers, with

$$r_1(1) = 6, \ r_1(2) = -3, \ r_1(3) = 0$$
$$r_2(1) = -2, \ r_2(2) = 0, \ r_2(3) = 6$$
$$r_3(1) = 10, \ r_3(2) = 10, \ r_3(3) = x$$

what must x equal if there is no arbitrage? For both parts, assume that you can simultaneously place wagers at any desired levels.

Exercise 6.4 Suppose, in Exercise 6.1, that one may also choose any pair of outcomes $i \neq j$ and bet that the outcome will be either i or j. What should the odds be on these three bets if an arbitrage opportunity is to be avoided?

Exercise 6.5 In Example 6.1a, show that if

$$\sum_{i=1}^{m} \frac{1}{1+o_i} \neq 1$$

then the betting scheme

$$x_i = \frac{(1+o_i)^{-1}}{1 - \sum_{i=1}^{m}(1+o_i)^{-1}}, \quad i = 1, \ldots, m,$$

will always yield a gain of exactly 1.

Exercise 6.6 In Example 6.1b, suppose one also has the option of purchasing a put option that allows its holder to put the stock for sale at the end of one period for a price of 150. Determine the value of P, the cost of the put, if there is to be no arbitrage; then show that the resulting call and put prices satisfy the put–call option parity formula (Proposition 5.2.2).

Exercise 6.7 Suppose that, in each period, the cost of a security either goes up by a factor of 2 or goes down by a factor of $1/2$ (i.e., $u = 2, d = 1/2$). If the initial price of the security is 100, determine the no-arbitrage cost of a call option to purchase the security at the end of two periods for a price of 150.

Exercise 6.8 Suppose, in Example 6.1b, that there are three possible prices for the security at time 1: 50, 100, or 200. (That is, allow for the possibility that the security's price remains unchanged.) Use the arbitrage theorem to find an interval for which there is no arbitrage if C lies in that interval.

A betting strategy $\mathbf{x}$ such that (using the notation of Section 6.1)

$$\sum_{i=1}^{n} x_i r_i(j) \geq 0, \quad j = 1, \ldots, m,$$

with strict inequality for at least one j, is said to be a *weak arbitrage* strategy. That is, whereas an arbitrage is present if there is a strategy that results in a positive gain for every outcome, a weak arbitrage is present if there is a strategy that never results in a loss and results in a positive gain for at least one outcome. (An arbitrage can be thought of as a *free lunch*, whereas a weak arbitrage is a *free lottery ticket*.) It can be shown that there will be no weak arbitrage if and only if there is a probability vector **p**, all of whose components are positive, such that

$$\sum_{j=1}^{m} p_j r_i(j) = 0, \quad i = 1, \ldots, n.$$

In other words, there will be no weak arbitrage if there is a probability vector that gives positive weight to each possible outcome and makes all bets fair.

Exercise 6.9 In Exercise 6.8, show that a weak arbitrage is possible if the cost of the option is equal to either endpoint of the interval determined.

Exercise 6.10 For the model of Section 6.2 with $n = 1$, show how an option can be replicated by a combination of borrowing and buying the security.

Exercise 6.11 The price of a security in each time period is its price in the previous time period multiplied either by $u = 1.25$ or by $d = .8$. The initial price of the security is 100. Consider the following "exotic" European call option that expires after five periods and has a strike price of 100. What makes this option exotic is that it becomes alive only if the price after two periods is strictly less than 100. That is, it becomes alive only if the price decreases in the first two periods. The final payoff of this option is

$$\text{payoff at time } 5 = I(S(5) - 100)^+,$$

where $I = 1$ if $S(2) < 100$ and $I = 0$ if $S(2) \geq 100$. Suppose the interest rate per period is $r = .1$.

(a) What is the no-arbitrage cost (at time 0) of this option?
(b) Is the cost of part (a) unique? Briefly explain.

(c) If each price change is equally likely to be an up or a down movement, what is the expected amount that an option holder receives at the time of expiration?

Exercise 6.12 Suppose the price of a security changes from period to period in such a manner that the price during period i is the price during period $i - 1$ multiplied either by $u = 1.1$ or by $d = 1/u$, $i \geq 1$. Suppose the price of the security in period 0 is 50. Aside from buying and selling the security, suppose one can also pay C in period 0 and receive either 100 in period 3 if the price in period 3 is at least 52, or 0 in period 3 if the price in that period is less than 52. Assuming an interest rate of $r = 0.05$, determine C if no arbitrage is possible.

REFERENCES

[1] De Finetti, Bruno (1937). "La prevision: ses lois logiques, ses sources subjectives." *Annales de l'Institut Henri Poincaré* 7: 1–68; English translation in S. Kyburg (Ed.) (1962), *Studies in Subjective Probability*, pp. 93–158. New York: Wiley.

[2] Gale, David (1960). *The Theory of Linear Economic Models*. New York: McGraw-Hill.

7. The Black–Scholes Formula

7.1 Introduction

In this chapter we derive the celebrated Black–Scholes formula, which gives – under the assumption that the price of a security evolves according to a geometric Brownian motion – the unique no-arbitrage cost of a call option on this security. Section 7.2 gives the derivation of the no-arbitrage cost, which is a function of five variables, and Section 7.3 discusses some of the properties of this function. Section 7.4 gives the strategy that can, in theory, be used to obtain an abitrage when the cost of the security is not as specified by the formula. Section 7.5, which is more theoretical than other sections of the text, presents simplified derivations of (1) the computational form of the Black–Scholes formula and (2) the partial derivatives of the no-arbitrage cost with respect to each of its five parameters.

7.2 The Black–Scholes Formula

Consider a call option having strike price K and expiration time t. That is, the option allows one to purchase a single unit of an underlying security at time t for the price K. Suppose further that the nominal interest rate is r, compounded continuously, and also that the price of the security follows a geometric Brownian motion with drift parameter μ and volatility parameter σ. Under these assumptions, we will find the unique cost of the option that does not give rise to an arbitrage.

To begin, let $S(y)$ denote the price of the security at time y. Because $\{S(y), 0 \leq y \leq t\}$ follows a geometric Brownian motion with volatility parameter σ and drift parameter μ, the n-stage approximation of this model supposes that, every t/n time units, the price changes; its new value is equal to its old value multiplied either by the factor

$$u = e^{\sigma\sqrt{t/n}} \text{ with probability } \frac{1}{2}\left(1 + \frac{\mu}{\sigma}\sqrt{t/n}\right)$$

or by the factor

$$d = e^{-\sigma\sqrt{t/n}} \quad \text{with probability} \quad \frac{1}{2}\left(1 - \frac{\mu}{\sigma}\sqrt{t/n}\right).$$

Thus, the n-stage approximation model is an n-stage binomial model in which the price at each time interval t/n either goes up by a multiplicative factor u or down by a multiplicative factor d. Therefore, if we let

$$X_i = \begin{cases} 1 & \text{if } S(it/n) = uS((i-1)t/n), \\ 0 & \text{if } S(it/n) = dS((i-1)t/n), \end{cases}$$

then it follows from the results of Section 6.2 that the only probability law on $X_1, \ldots, X_n$ that makes all security buying bets fair in the n-stage approximation model is the one that takes the X_i to be independent with

$$p \equiv P\{X_i = 1\}$$

$$= \frac{1 + rt/n - d}{u - d}$$

$$= \frac{1 - e^{-\sigma\sqrt{t/n}} + rt/n}{e^{\sigma\sqrt{t/n}} - e^{-\sigma\sqrt{t/n}}}.$$

Using the first three terms of the Taylor series expansion about 0 of the function e^x shows that

$$e^{-\sigma\sqrt{t/n}} \approx 1 - \sigma\sqrt{t/n} + \sigma^2 t/2n,$$

$$e^{\sigma\sqrt{t/n}} \approx 1 + \sigma\sqrt{t/n} + \sigma^2 t/2n.$$

Therefore,

$$p \approx \frac{\sigma\sqrt{t/n} - \sigma^2 t/2n + rt/n}{2\sigma\sqrt{t/n}}$$

$$= \frac{1}{2} + \frac{r\sqrt{t/n}}{2\sigma} - \frac{\sigma\sqrt{t/n}}{4}$$

$$= \frac{1}{2}\left(1 + \frac{r - \sigma^2/2}{\sigma}\sqrt{t/n}\right).$$

That is, the unique risk-neutral probabilities on the n-stage approximation model result from supposing that, in each period, the price either goes up by the factor $e^{\sigma\sqrt{t/n}}$ with probability p or goes down by the factor $e^{-\sigma\sqrt{t/n}}$ with probability $1 - p$. But, from Section 3.2, it follows that as $n \to \infty$ this risk-neutral probability law converges to geometric Brownian motion with drift coefficient $r - \sigma^2/2$ and volatility parameter σ. Because the n-stage approximation model becomes the geometric Brownian motion as n becomes larger, it is reasonable to suppose (and can be rigorously proven) that this risk-neutral geometric Brownian motion is the only probability law on the evolution of prices over time that makes all security buying bets fair. (In other words, we have just argued that if the underlying price of a security follows a geometric Brownian motion with volatility parameter σ, then the only probability law on the sequence of prices that results in all security buying bets being fair is that of a geometric Brownian motion with drift parameter $r - \sigma^2/2$ and volatility parameter σ.) Consequently, by the arbitrage theorem, either options are priced to be fair bets according to the risk-neutral geometric Brownian motion probability law or else there will be an arbitrage.

Now, under the risk-neutral geometric Brownian motion, $S(t)/S(0)$ is a lognormal random variable with mean parameter $(r - \sigma^2/2)t$ and variance parameter $\sigma^2 t$. Hence C, the unique no-arbitrage cost of a call option to purchase the security at time t for the specified price K, is

$$C = e^{-rt}E[(S(t) - K)^+]$$

$$= e^{-rt}E[(S(0)e^W - K)^+], \tag{7.1}$$

where W is a normal random variable with mean $(r - \sigma^2/2)t$ and variance $\sigma^2 t$.

The right side of Equation (7.1) can be explicitly evaluated (see Section 7.4 for the derivation) to give the following expression, known as the *Black–Scholes option pricing formula*:

$$C = S(0)\Phi(\omega) - Ke^{-rt}\Phi(\omega - \sigma\sqrt{t}), \tag{7.2}$$

where

$$\omega = \frac{rt + \sigma^2 t/2 - \log(K/S(0))}{\sigma\sqrt{t}}$$

and where $\Phi(x)$ is the standard normal distribution function.

Example 7.1a Suppose that a security is presently selling for a price of 30, the nominal interest rate is 8% (with the unit of time being one year), and the security's volatility is .20. Find the no-arbitrage cost of a call option that expires in three months and has a strike price of 34.

Solution. The parameters are

$$t = .25, \quad r = .08, \quad \sigma = .20, \quad K = 34, \quad S(0) = 30,$$

so we have

$$\omega = \frac{.02 + .005 - \log(34/30)}{(.2)(.5)} \approx -1.0016.$$

Therefore,

$$C = 30\Phi(-1.0016) - 34e^{-.02}\Phi(-1.1016)$$

$$= 30(.15827) - 34(.9802)(.13532)$$

$$\approx .2383.$$

The appropriate price of the option is thus 24 cents.   □

Remarks. 1. Another way to derive the no-arbitrage option cost C is to consider the unique no-arbitrage cost of an option in the n-period approximation model and then let n go to infinity.

2. Let $C(s, t, K)$ be the no-arbitrage cost of an option having strike price K and exercise time t when the initial price of the security is s. That is, $C(s, t, K)$ is the C of the Black–Scholes formula having $S(0) = s$. If the price of the underlying security at time y ($0 < y < t$) is $S(y) = s_y$, then $C(s_y, t - y, K)$ is the unique no-arbitrage cost of the option at time y. This is because, at time y, the option will expire after an additional time $t - y$ with the same exercise price K, and for the next $t - y$ units of time the security will follow a geometric Brownian motion with initial value s_y.

3. It follows from the put–call option parity formula given in Proposition 5.2.2 that the no-arbitrage cost of a European put option with initial price s, strike price K, and exercise time t – call it $P(s, t, K)$ – is given by

$$P(s, t, K) = C(s, t, K) + Ke^{-rt} - s,$$

where $C(s, t, K)$ is the no-arbitrage cost of a call option on the same stock.

4. Because the risk-neutral geometric Brownian motion depends only on σ and not on μ, it follows that the no-arbitrage cost of the option depends on the underlying Brownian motion only through its volatility parameter σ and not its drift parameter.

5. The no-arbitrage option cost is unchanged if the security's price over time is assumed to follow a geometric Brownian motion with a fixed volatility σ but with a drift that varies over time. Because the n-stage approximation model for the price history up to time t of the time-varying drift process is still a binomial up–down model with $u = e^{\sigma\sqrt{t/n}}$ and $d = e^{-\sigma\sqrt{t/n}}$, it has the same unique risk-neutral probability law as when the drift parameter is unchanging, and thus it will give rise to the same unique no-arbitrage option cost. (The only way that a changing drift parameter would affect our derivation of the Black–Scholes formula is by leading to different probabilities for up moves in the different time periods, but these probabilities have no effect on the the risk-neutral probabilities.)

7.3   Properties of the Black–Scholes Option Cost

The no-arbitrage option cost $C = C(s, t, K, \sigma, r)$ is a function of five variables: the security's initial price s; the expiration time t of the option; the strike price K; the security's volatility parameter σ; and the interest rate r. To see what happens to the cost as a function of each of these variables, we use Equation (7.1):

$$C(s, t, K, \sigma, r) = e^{-rt}E[(se^W - K)^+],$$

where W is a normal random variable with mean $(r - \sigma^2/2)t$ and variance $\sigma^2 t$.

Properties of $C = C(s, t, K, \sigma, r)$

1. *C is an increasing, convex function of s.*

This means that if the other four variables remain the same, then the no-arbitrage cost of the option is an increasing function of the security's initial price as well as a convex function of the security's initial price. These results (the first of which is very intuitive) follow from Equation (7.1). To see why, first note (see Figure 7.1) that, for any positive constant a, the function $e^{-rt}(sa - K)^+$ is an increasing, convex function

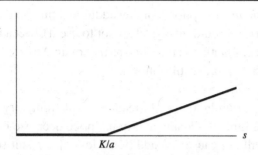

Figure 7.1: The Increasing, Convex Function $f(s) = e^{-rt}(sa - K)^+$

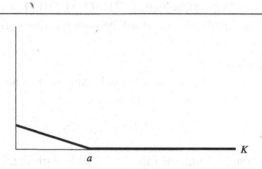

Figure 7.2: The Decreasing, Convex Function $f(K) = e^{-rt}(a - K)^+$

of s. Consequently, because the probability distribution of W does not depend on s, the quantity $e^{-rt}(se^W - K)^+$ is, for all W, increasing and convex in s, and thus so is its expected value.

2. *C is a decreasing, convex function of K.*

This follows from the fact that $e^{-rt}(se^W - K)^+$ is, for all W, decreasing and convex in K (see Figure 7.2), and thus so is its expectation. (This is in agreement with the more general arbitrage argument made in Section 5.2, which did not assume a model for the security's price evolution.)

3. *C is increasing in t.*

Although a mathematical argument can be given (see Section 7.4), a simpler and more intuitive argument is obtained by noting that it is immediate that the option cost would be increasing in t if the option were an *American* call option (for any additional time to exercise could not hurt, since one could always elect not to use it). Because the value of a European call option is the same as that of an American call option (Proposition 5.2.1), the result follows.

4. *C is increasing in σ.*

Because an option holder will greatly benefit from very large prices at the exercise time, while any additional price decrease below the exercise price will not cause any additional loss, this result seems at first sight to be quite intuitive. However, it is more subtle than it appears, because an increase in σ results not only in an increase in the variance of the logarithm of the final price under the risk-neutral valuation but also in a *decrease* in the mean (since $E[\log(S(t)/S(0))] = (r - \sigma^2/2)t)$. Nevertheless, the result is true and will be shown mathematically in Section 7.4.

5. *C is increasing in r.*

To verify this property, note that we can express W, a normal random variable with mean $(r - \sigma^2/2)t$ and variance $t\sigma^2$, as

$$W = rt - \sigma^2 t/2 + \sigma\sqrt{t}Z,$$

where Z is a standard normal random variable with mean 0 and variance 1. Hence, from Equation (7.1) we have that

$$C = E[(se^{-\sigma^2 t/2 + \sigma\sqrt{t}Z} - Ke^{-rt})^+].$$

The result now follows because $(se^{-\sigma^2 t/2 + \sigma\sqrt{t}Z} - Ke^{-rt})^+$, and thus its expected value, is increasing in r. Indeed, it follows from the preceding that, under the no-arbitrage geometric Brownian motion model, the only effect of an increased interest rate is that it reduces the present value of the amount to be paid if the option is exercised, thus increasing the value of the option.

The rate of change in the value of the call option as a function of a change in the price of the underlying security is described by the quantity *delta*, denoted as Δ. Formally, if $C(s, t, K, \sigma, r)$ is the Black–Scholes cost valuation of the option, then Δ is its partial derivative with respect to s; that is,

$$\Delta = \frac{\partial}{\partial s}C(s, t, K, \sigma, r).$$

In Section 7.4 we will show that

$$\Delta = \Phi(\omega)$$

where, as given in Equation (7.2),

$$\omega = \frac{rt + \sigma^2 t/2 - \log(K/S(0))}{\sigma\sqrt{t}}.$$

Delta can be used to construct investment portfolios that hedge against risk. For instance, suppose that an investor feels that a call option is underpriced and consequently buys the call. To protect himself against a decrease in its price, he can simultaneously sell a certain number of shares of the security. To determine how many shares he should sell, note that if the price of the security decreases by the small amount h then the worth of the option will decrease by the amount $h\Delta$, implying that the investor would be covered if he sold Δ shares of the security. Therefore, a reasonable hedge might be to sell Δ shares of the security for each option purchased. This heuristic argument will be made precise in the next section, where we present the delta hedging arbitrage strategy – a strategy that can, in theory, be used to construct an arbitrage if a call option is not priced according to the Black–Scholes formula.

7.4   The Delta Hedging Arbitrage Strategy

In this section we show how the payoff from an option can be replicated by a fixed initial payment (divided into an initial purchasing of shares and an initial bank deposit, where either might be negative) and a continual readjustment of funds. We first present it for the finite-stage approximation model and then for the geometric Brownian motion model for the security's price evolution.

To begin, consider a security whose initial price is s and suppose that, after each time period, its price changes either by the multiple u or by the multiple d. Let us determine the amount of money x that you must have at time 0 in order to meet a payment, at time 1, of a if the price of the stock is us at time 1 or of b if the price at time 1 is ds. To determine x, and the investment that enables you to meet the payment, suppose that you purchase y shares of the stock and then either put the remaining $x - ys$ in the bank if $x - ys \geq 0$ or borrow $ys - x$ from the bank if $x - ys < 0$. Then, for the initial cost of x, you will have a return at time 1 given by

$$\text{return at time 1} = \begin{cases} yus + (x - ys)(1 + r) & \text{if } S(1) = us, \\ yds + (x - ys)(1 + r) & \text{if } S(1) = ds, \end{cases}$$

where $S(1)$ is the price of the security at time 1 and r is the interest rate per period. Thus, if we choose x and y such that

$$yus + (x - ys)(1 + r) = a,$$

$$yds + (x - ys)(1 + r) = b,$$

then after taking our money out of the bank (or meeting our loan payment) we will have the desired amount. Subtracting the second equation from the first gives that

$$y = \frac{a - b}{s(u - d)}.$$

Substituting the preceding expression for y into the first equation yields

$$\frac{a - b}{u - d}[u - (1 + r)] + x(1 + r) = a$$

or

$$\begin{aligned} x &= \frac{1}{1 + r}\left(a\left[1 - \frac{u - (1 + r)}{u - d}\right] + b\frac{u - 1 - r}{u - d}\right) \\ &= \frac{1}{1 + r}\left(a\frac{1 + r - d}{u - d} + b\frac{u - 1 - r}{u - d}\right) \\ &= p\frac{a}{1 + r} + (1 - p)\frac{b}{1 + r}, \end{aligned}$$

where

$$p = \frac{1 + r - d}{u - d}.$$

In other words, the amount of money that is needed at time 0 is equal to the expected present value, under the risk-neutral probabilities, of the payoff at time 1. Moreover, the investment strategy calls for purchasing of $y = \frac{a-b}{s(u-d)}$ shares of the security and putting the remainder in the bank.

Remark. If $a > b$, as it would be if the payoff at time 1 results from paying the holder of a call option, then $y > 0$ and so a positive amount of the security is purchased; if $a < b$, as it would be if the payoff at time 1 results from paying the holder of a put option, then $y < 0$ and so $-y$ shares of the security are sold short.

Now consider the problem of determining how much money is needed at time 0 to meet a payoff at time 2 of $x_{i,2}$ if the price of the security at time 2 is $u^i d^{2-i} s$ ($i = 0, 1, 2$). To solve this problem, let us first determine, for each possible price of the security at time 1, the amount that is needed at time 1 to meet the payment at time 2. If the price at time 1 is us, then the amount needed at time 2 would be either $x_{2,2}$ if the price at time 2 is $u^2 s$ or $x_{1,2}$ if the price is uds. Thus, it follows from our preceding analysis that if the price at time 1 is us then we would, at time 1, need the amount

$$x_{1,1} = p\frac{x_{2,2}}{1+r} + (1-p)\frac{x_{1,2}}{1+r},$$

and the strategy is to purchase

$$y_{1,1} = \frac{x_{2,2} - x_{1,2}}{us(u-d)}$$

shares of the security and put the remainder in the bank. Similarly, if the price at time 1 is ds, then to meet the final payment at time 2 we would, at time 1, need the amount

$$x_{0,1} = p\frac{x_{1,2}}{1+r} + (1-p)\frac{x_{0,2}}{1+r},$$

and the strategy is to purchase

$$y_{0,1} = \frac{x_{1,2} - x_{0,2}}{ds(u-d)}$$

shares of the security and put the remainder in the bank. Now, at time 0 we need to have enough to invest so as to be able to have either $x_{1,1}$ or $x_{0,1}$ at time 1, depending on whether the price of the security is us or ds at that time. Consequently, at time 0 we need the amount

$$x_{0,0} = p\frac{x_{1,1}}{1+r} + (1-p)\frac{x_{0,1}}{1+r}$$

$$= p^2\frac{x_{2,2}}{(1+r)^2} + 2p(1-p)\frac{x_{1,2}}{(1+r)^2} + (1-p)^2\frac{x_{0,2}}{(1+r)^2}.$$

That is, once again the amount needed is the expected present value, under the risk-neutral probabilities, of the final payoff. The strategy is to purchase

$$y_{0,0} = \frac{x_{1,1} - x_{0,1}}{s(u - d)}$$

shares of the security and put the remainder in the bank.

The preceding is easily generalized to an n-period problem, where the payoff at the end of period n is $x_{i,n}$ if the price at that time is $u^i d^{n-i} s$. The amount $x_{i,j}$ needed at time j, given that the price of the security at that time is $u^i d^{j-i} s$, is equal to the conditional expected time-j value of the final payoff, where the expected value is computed under the assumption that the successive changes in price are governed by the risk-neutral probabilities. (That is, the successive changes are independent, with each new price equal to the previous period's price multiplied either by the factor u with probability p or by the factor d with probability $1 - p$.)

If the payoff results from paying the holder of a call option that has strike price K and expiration time n, then the payoff at time n is

$$x_{i,n} = (u^i d^{n-i} s - K)^+, \quad i = 0, \ldots, n,$$

when the price of the security at time n is $u^i d^{n-i} s$. Because our investment strategy replicates the payoff from this option, it follows from the law of one price (as well as from the arbitrage theorem) that $x_{0,0}$, the initial amount needed, is equal to the unique no-arbitrage cost of the option. Moreover, $x_{i,j}$, the amount needed at time j when the price at that time is $su^i d^{j-i}$, is the unique no-arbitrage cost of the option at that time and price. To effect an arbitrage when C, the cost of the option at time 0, is larger than $x_{0,0}$, we can sell the option, use $x_{0,0}$ from this sale to meet the option payoff at time n, and walk away with a positive profit of $C - x_{0,0}$. Now, suppose that $C < x_{0,0}$. Because the investment procedure we developed transforms an initial fortune of $x_{0,0}$ into a time-n fortune of $x_{i,n}$ if the price of the security at that time is $u^i d^{n-i} s$ ($i = 0, \ldots, n$), it follows that by reversing the procedure (changing buying into selling, and vice versa) we can transform an initial debt of $x_{0,0}$ into a time-n debt of $x_{i,n}$ when the price at time n is $su^i d^{n-i}$. Consequently, when $C < x_{0,0}$, we can make an arbitrage by borrowing the amount $x_{0,0}$, using C of this amount to buy the option, and then using the investment procedure to transform the initial debt into a time-n debt whose amount is exactly that of the return from the option. Hence, in either case we can gain $|C - x_{0,0}|$ at time 0; we then follow an investment strategy that guarantees we have no additional losses or gains. In other words, after taking our profit, our strategy *hedges* all future risks.

Let us now determine the hedging strategy for a call option with strike price K when the price of the security follows a geometric Brownian motion with volatility σ. To begin, consider the finite-period approximation, where each h time units the price of the security either increases by the factor $e^{\sigma\sqrt{h}}$ or decreases by the factor $e^{-\sigma\sqrt{h}}$. Suppose the present price of the stock is s and the call option expires after an additional time t. Because the price after an additional time h is either $se^{\sigma\sqrt{h}}$ or $se^{-\sigma\sqrt{h}}$, it follows that the amount we will need in the next period to utilize the hedging strategy is either $C(se^{\sigma\sqrt{h}}, t - h)$ if the price is $se^{\sigma\sqrt{h}}$ or $C(se^{-\sigma\sqrt{h}}, t - h)$ if the price is $se^{-\sigma\sqrt{h}}$, where $C(s, t)$ is the no-arbitrage cost of the call option with strike price K when the current price of the security is s and the option expires after an additional time t. (This notation suppresses the dependence of C on K, r, and σ.) Consequently, when the price of the security is s and time t remains before the option expires, the hedging strategy calls for owning

$$\frac{C(se^{\sigma\sqrt{h}}, t - h) - C(se^{-\sigma\sqrt{h}}, t - h)}{se^{\sigma\sqrt{h}} - se^{-\sigma\sqrt{h}}}$$

shares of the security.

To determine, under geometric Brownian motion, the number of shares of the security that should be owned when the price of the security is s and the call option expires after an additional time t, we need to let h go to zero in the preceding expression. Thus, we need to determine

$$\lim_{h \to 0} \frac{C(se^{\sigma\sqrt{h}}, t - h) - C(se^{-\sigma\sqrt{h}}, t - h)}{se^{\sigma\sqrt{h}} - se^{-\sigma\sqrt{h}}}$$

$$= \lim_{a \to 0} \frac{C(se^{\sigma a}, t - a^2) - C(se^{-\sigma a}, t - a^2)}{se^{\sigma a} - se^{-\sigma a}}.$$

However, calculus (L'Hôpital's rule along with the chain rule for differentiating a function of two variables) yields

$$\lim_{a\to 0} \frac{C(se^{\sigma a}, t - a^2) - C(se^{-\sigma a}, t - a^2)}{se^{\sigma a} - se^{-\sigma a}}$$

$$= \lim_{a\to 0} \frac{s\sigma e^{\sigma a}\frac{\partial}{\partial y}C(y, t)|_{y=se^{\sigma a}} + s\sigma e^{-\sigma a}\frac{\partial}{\partial y}C(y, t)|_{y=se^{-\sigma a}}}{s\sigma e^{\sigma a} + s\sigma e^{-\sigma a}}$$

$$= \frac{\partial}{\partial y}C(y, t)|_{y=s}$$

$$= \frac{\partial}{\partial s}C(s, t).$$

Therefore, the return from a call option having strike price K and exercise time T can be replicated by an investment strategy that requires an investment capital of $C(S(0), T, K)$ and then calls for owning exactly $\frac{\partial}{\partial s}C(s, t, K)$ shares of the security when its current price is s and time t remains before the option expires, with the absolute value of your remaining capital at that time being either in the bank (if your remaining capital is positive) or borrowed (if it is negative).

Suppose the market price of the (K, T) call option is greater than $C(S(0), T, K)$; then an arbitrage can be made by selling the option and using $C(S(0), T, K)$ from this sale along with the preceding strategy to replicate the return from the option. When the market cost C is less than $C(S(0), T, K)$, an arbitrage is obtained by doing the reverse. Namely, borrow $C(S(0), T, K)$ and use C of this amount to buy a (K, T) call option (what remains will be yours to keep); then maintain a short position of $\frac{\partial}{\partial s}C(s, t, K)$ shares of the security when its current price is s and time t remains before the option expires. The invested money from these short positions, along with your call option, will cover your loan of $C(S(0), T, K)$ and also pay off your final short position.

7.5   Some Derivations

In Section 7.5.1 we give the derivation of Equation (7.2), the computational form of the Black–Scholes formula. In Section 7.5.2 we derive the partial derivative of $C(s, t, K, \sigma, r)$ with respect to each of the quantities s, t, K, σ, and r.

7.5.1   *The Black–Scholes Formula*

Let

$$C(s, t, K, \sigma, r) = E[e^{-rt}(S(t) - K)^+]$$

be the risk-neutral cost of a call option with strike price K and expiration time t when the interest rate is r and the underlying security, whose initial price is s, follows a geometric Brownian motion with volatility parameter σ. To derive the Black–Scholes option pricing formula as well as the partial derivatives of C, we will use the fact that, under the risk-neutral probabilities, $S(t)$ can be expressed as

$$S(t) = s \exp\{(r - \sigma^2/2)t + \sigma\sqrt{t}Z\}, \tag{7.3}$$

where Z is a standard normal random variable.

Let I be the indicator random variable for the event that the option finishes in the money. That is,

$$I = \begin{cases} 1 & \text{if } S(t) > K, \\ 0 & \text{if } S(t) \leq K. \end{cases} \tag{7.4}$$

We will use the following lemmas.

Lemma 7.5.1   *Using the representations* (7.3) *and* (7.4),

$$I = \begin{cases} 1 & \text{if } Z > \sigma\sqrt{t} - \omega, \\ 0 & \text{otherwise,} \end{cases}$$

where

$$\omega = \frac{rt + \sigma^2 t/2 - \log(K/s)}{\sigma\sqrt{t}}.$$

Proof.

$$S(t) > K \iff \exp\{(r - \sigma^2/2)t + \sigma\sqrt{t}Z\} > K/s$$
$$\iff Z > \frac{\log(K/s) - (r - \sigma^2/2)t}{\sigma\sqrt{t}}$$
$$\iff Z > \sigma\sqrt{t} - \omega. \qquad \square$$

Lemma 7.5.2

$$E[I] = P\{S(t) > K\} = \Phi(\omega - \sigma\sqrt{t}),$$

where Φ is the standard normal distribution function.

Proof. It follows from its definition that

$$
\begin{aligned}
E[I] &= P\{S(t) > K\} \\
&= P\{Z > \sigma\sqrt{t} - \omega\} \quad \text{(from Lemma 7.5.1)} \\
&= P\{Z < \omega - \sigma\sqrt{t}\} \\
&= \Phi(\omega - \sigma\sqrt{t}).
\end{aligned}
$$
$\qquad\square$

Lemma 7.5.3

$$e^{-rt}E[IS(t)] = s\Phi(\omega).$$

Proof. With $c = \sigma\sqrt{t} - \omega$, it follows from the representation (7.3) and Lemma 7.5.1 that

$$
\begin{aligned}
E[IS(t)] &= \int_c^\infty s \exp\{(r - \sigma^2/2)t + \sigma\sqrt{t}x\}\frac{1}{\sqrt{2\pi}}e^{-x^2/2}\,dx \\
&= \frac{1}{\sqrt{2\pi}}s \exp\{(r - \sigma^2/2)t\}\int_c^\infty \exp\{-(x^2 - 2\sigma\sqrt{t}x)/2\}\,dx \\
&= \frac{1}{\sqrt{2\pi}}se^{rt} \int_c^\infty \exp\{-(x - \sigma\sqrt{t})^2/2\}\,dx \\
&= se^{rt}\frac{1}{\sqrt{2\pi}}\int_{-\omega}^\infty e^{-y^2/2}\,dy \quad \text{(by letting } y = x - \sigma\sqrt{t}) \\
&= se^{rt}P\{Z > -\omega\} \\
&= se^{rt}\Phi(\omega).
\end{aligned}
$$
$\qquad\square$

Theorem 7.5.1 (The Black–Scholes Pricing Formula)

$$C(s, t, K, \sigma, r) = s\Phi(\omega) - Ke^{-rt}\Phi(\omega - \sigma\sqrt{t}).$$

Proof.

$$C(s, t, K, \sigma, r) = e^{-rt}E[(S(t) - K)^+]$$
$$= e^{-rt}E[I(S(t) - K)]$$
$$= e^{-rt}E[I(S(t)] - Ke^{-rt}E[I],$$

and the result follows from Lemmas 7.5.2 and 7.5.3. $\qquad\square$

7.5.2 The Partial Derivatives

Let Z be a normal random variable with mean 0 and variance 1, and let $W = (r - \sigma^2/2)t + \sigma\sqrt{t}Z$. Thus, W is normal with mean $(r - \sigma^2/2)t$ and variance $t\sigma^2$.

The Black-Scholes call option formula can be written as

$$C = C(s, t, K, \sigma, r) = E[e^{-rt}I(se^W - K)]$$

where
$$I = \begin{cases} 1, & \text{if } se^W > K \\ 0, & \text{if } se^W \le K \end{cases}$$

is the indicator of the event that $se^W > K$. Now,

$$e^{-rt}I(se^W - K) = \begin{cases} e^{-rt}(se^W - K), & \text{if } se^W > K \\ 0, & \text{if } se^W \le K \end{cases}$$

As the preceding is, for given Z, a differentiable function of the parameters s, t, K, σ, r, we see that for x equal to any one of these variables,

$$\frac{\partial}{\partial x}e^{-rt}I(se^W - K) = \begin{cases} \frac{\partial}{\partial x}e^{-rt}(se^W - K), & \text{if } se^W > K \\ 0, & \text{if } se^W \le K \end{cases}$$

That is,
$$\frac{\partial}{\partial x}e^{-rt}I(se^W - K) = I\frac{\partial}{\partial x}e^{-rt}(se^W - K)$$

Using that the partial derivative and the expectation operation can be interchanged, the preceding gives that

$$\frac{\partial C}{\partial x} = \frac{\partial}{\partial x}E\left[e^{-rt}I\left(se^W - K\right)\right]$$
$$= E\left[\frac{\partial}{\partial x}e^{-rt}I\left(se^W - K\right)\right]$$
$$= E\left[I\frac{\partial}{\partial x}e^{-rt}\left(se^W - K\right)\right] \qquad (7.5)$$

We will now derive the partial derivatives of C with respect to K, s, and r.

Proposition 7.5.1

$$\frac{\partial C}{\partial K} = -e^{-rt}\Phi(\omega - \sigma\sqrt{t}).$$

Proof. Because $S(t)$ does not depend on K,

$$\frac{\partial}{\partial K}e^{-rt}(S(t) - K) = -e^{-rt}.$$

Using Equation (7.5), this gives

$$\frac{\partial C}{\partial K} = E[-Ie^{-rt}]$$
$$= -e^{-rt}E[I]$$
$$= -e^{-rt}\Phi(\omega - \sigma\sqrt{t}),$$

where the final equality used Lemma 7.5.2.   □

As noted previously, $\frac{\partial C}{\partial s}$ is called *delta*.

Proposition 7.5.2

$$\frac{\partial C}{\partial s} = \Phi(\omega).$$

Proof. Using the representation of Equation (7.3), we see that

$$\frac{\partial}{\partial s}e^{-rt}(S(t) - K) = e^{-rt}\frac{\partial S(t)}{\partial s} = \frac{S(t)}{s}e^{-rt}.$$

Hence, by Equation (7.5),

$$\frac{\partial C}{\partial s} = \frac{e^{-rt}}{s}E[IS(t)]$$
$$= \Phi(\omega),$$

where the final equality used Lemma 7.5.3.   □

The partial derivative of C with respect to r is called *rho*.

Proposition 7.5.3

$$\frac{\partial C}{\partial r} = Kte^{-rt}\Phi(\omega - \sigma\sqrt{t}).$$

Proof.

$$\frac{\partial}{\partial r}[e^{-rt}(S(t) - K)] = -te^{-rt}(S(t) - K) + e^{-rt}\frac{\partial S(t)}{\partial r}$$

$$= -te^{-rt}(S(t) - K) + e^{-rt}tS(t) \qquad \text{(from (7.3))}$$

$$= Kte^{-rt}.$$

Therefore, by Equation (7.5) and Lemma 7.5.2,

$$\frac{\partial C}{\partial r} = Kte^{-rt}E[I] = Kte^{-rt}\Phi(\omega - \sigma\sqrt{t}). \qquad \square$$

In order to determine the other partial derivatives, we need an additional lemma, whose proof is similar to that of Lemma 7.5.3.

Lemma 7.5.4 *With $S(t)$ as given by Equation (7.3),*

$$e^{-rt}E[IS(t)Z] = s(\Phi'(\omega) + \sigma\sqrt{t}\Phi(\omega)).$$

Proof. With $c = \sigma\sqrt{t} - \omega$, it follows from Lemma 7.5.1 that

$$E[IZS(t)]$$

$$= \int_c^\infty xs\exp\{(r - \sigma^2/2)t + \sigma\sqrt{t}x\}\frac{1}{\sqrt{2\pi}}e^{-x^2/2}\,dx$$

$$= \frac{1}{\sqrt{2\pi}}s\exp\{(r - \sigma^2/2)t\}\int_c^\infty x\exp\{-(x^2 - 2\sigma\sqrt{t}x)/2\}\,dx$$

$$= \frac{1}{\sqrt{2\pi}}se^{rt}\int_c^\infty x\exp\{-(x - \sigma\sqrt{t})^2/2\}\,dx$$

$$= \frac{1}{\sqrt{2\pi}}se^{rt}\int_{-\omega}^\infty (y + \sigma\sqrt{t})e^{-y^2/2}\,dy \qquad \text{(by letting } y = x - \sigma\sqrt{t})$$

$$= se^{rt}\left[\int_{-\omega}^\infty \frac{1}{\sqrt{2\pi}}ye^{-y^2/2}\,dy + \sigma\sqrt{t}\frac{1}{\sqrt{2\pi}}\int_{-\omega}^\infty e^{-y^2/2}\,dy\right]$$

$$= se^{rt}\left[\frac{1}{\sqrt{2\pi}}e^{-\omega^2/2} + \sigma\sqrt{t}\Phi(\omega)\right]. \qquad \square$$

The partial derivative of C with respect to σ is called *vega*.

Proposition 7.5.4

$$\frac{\partial C}{\partial \sigma} = s\sqrt{t}\,\Phi'(\omega).$$

Proof. Equation (7.3) yields that

$$\frac{\partial}{\partial \sigma}[e^{-rt}(S(t) - K)] = e^{-rt}S(t)(-t\sigma + \sqrt{t}Z).$$

Hence, by Equation (7.5),

$$\begin{aligned}
\frac{\partial C}{\partial \sigma} &= E[e^{-rt}IS(t)(-t\sigma + \sqrt{t}Z)] \\
&= -t\sigma e^{-rt}E[IS(t)] + \sqrt{t}e^{-rt}E[IS(t)Z] \\
&= -t\sigma s\Phi(\omega) + s\sqrt{t}(\Phi'(\omega) + \sigma\sqrt{t}\Phi(\omega)) \\
&= s\sqrt{t}\,\Phi'(\omega),
\end{aligned}$$

where the next-to-last equality used Lemmas 7.5.3 and 7.5.4.   □

The negative of the partial derivative of C with respect to t is called *theta*.

Proposition 7.5.5

$$\frac{\partial C}{\partial t} = \frac{\sigma}{2\sqrt{t}}s\Phi'(\omega) + Kre^{-rt}\Phi(\omega - \sigma\sqrt{t}).$$

Proof.

$$\begin{aligned}
\frac{\partial}{\partial t}[e^{-rt}(S(t) - K)] &= e^{-rt}\frac{\partial S(t)}{\partial t} - re^{-rt}S(t) + Kre^{-rt} \\
&= e^{-rt}S(t)\left(r - \frac{\sigma^2}{2} + \frac{\sigma}{2\sqrt{t}}Z\right) \\
&\quad - re^{-rt}S(t) + Kre^{-rt} \\
&= e^{-rt}S(t)\left(\frac{-\sigma^2}{2} + \frac{\sigma}{2\sqrt{t}}Z\right) + Kre^{-rt}.
\end{aligned}$$

Therefore, using Equation (7.5),

$$\frac{\partial C}{\partial t} = -e^{-rt}E[IS(t)]\frac{\sigma^2}{2} + e^{-rt}E[IZS(t)]\frac{\sigma}{2\sqrt{t}} + Kre^{-rt}E[I]$$

$$= -s\Phi(\omega)\frac{\sigma^2}{2} + \frac{\sigma}{2\sqrt{t}}s(\Phi'(\omega) + \sigma\sqrt{t}\Phi(\omega))$$

$$+ Kre^{-rt}\Phi(\omega - \sigma\sqrt{t})$$

$$= \frac{\sigma}{2\sqrt{t}}s\Phi'(\omega) + Kre^{-rt}\Phi(\omega - \sigma\sqrt{t}). \qquad \square$$

Remark. To calculate vega and theta, use that $\Phi'(x)$ is the standard normal density function given by

$$\Phi'(x) = \frac{1}{\sqrt{2\pi}}e^{-x^2/2}.$$

The following corollary uses the partial derivatives to present a more analytic proof of the results of Section 7.2.

Corollary 7.5.1 $C(s, t, K, \sigma, r)$ *is*

(a) *decreasing and convex in* K;
(b) *increasing and convex in* s;
(c) *increasing, but neither convex nor concave, in* r, σ, *and* t.

Proof. (a) From Proposition 7.5.1, we have $\frac{\partial C}{\partial K} < 0$, and

$$\frac{\partial^2 C}{\partial K^2} = -e^{-rt}\Phi'(\omega - \sigma\sqrt{t})\frac{\partial \omega}{\partial K}$$

$$= e^{-rt}\Phi'(\omega - \sigma\sqrt{t})\frac{1}{K\sigma\sqrt{t}}$$

$$> 0.$$

(b) It follows from Proposition 7.5.2 that $\frac{\partial C}{\partial s} > 0$, and

$$\frac{\partial^2 C}{\partial s^2} = \Phi'(\omega)\frac{\partial \omega}{\partial s}$$

$$= \Phi'(\omega)\frac{1}{s\sigma\sqrt{t}} \qquad (7.6)$$

$$> 0.$$

(c) It follows from Propositions 7.5.3, 7.5.4, and 7.5.5 that, for $x = r, \sigma, t$,

$$\frac{\partial C}{\partial x} > 0,$$

which proves the monotonicity. Because each of the second derivatives can be shown to be sometimes positive and sometimes negative, it follows that C is neither convex nor concave in r, σ, or t.   □

Remarks. The results that $C(s, t, K, \sigma, r)$ is decreasing and convex in K and increasing in t would be true no matter what model we assumed for the price evolution of the security. The results that $C(s, t, K, \sigma, r)$ is increasing and convex in s, increasing in r, and increasing in σ depend on the assumption that the price evolution follows a geometric Brownian motion with volatility parameter σ. The second partial derivative of C with respect to s, whose value is given by Equation (7.6), is called *gamma*.

7.6   European Put Options

The put call option parity formula, in conjunction with the Black-Scholes equation, yields the unique no arbitrage cost of a European (K, t) put option:

$$P(s, t, K, r, \sigma) = C(s, t, K, r, \sigma) + Ke^{-rt} - s \qquad (7.7)$$

Whereas the preceding is useful for computational purposes, to determine monotonicity and convexity properties of $P = P(s, t, K, r, \sigma)$ it is also useful to use that $P(s, t, K, r, \sigma)$ must equal the expected return from the put under the risk neutral geometric Brownian motion process. Consequently, with Z being a standard normal random variable,

$$P(s, t, K, r, \sigma) = e^{-rt} E[(K - se^{(r-\frac{\sigma^2}{2})t+\sigma\sqrt{t}Z})^+]$$

$$= E[(Ke^{-rt} - se^{-\frac{\sigma^2}{2}t+\sigma\sqrt{t}Z})^+]$$

Now, for a fixed value of Z, the function $(Ke^{-rt} - se^{-\frac{\sigma^2}{2}t+\sigma\sqrt{t}Z})^+$ is

1. Decreasing and convex in s. (This follows because $(a - bs)^+$ is, for $b > 0$, decreasing and convex in s.)

2. Decreasing and convex in r. (This follows because $(ae^{-rt} - b)^+$ is, for $a > 0$, decreasing and convex in r.)

3. Increasing and convex in K. (This follows because $(aK - b)^+$ is, for $a > 0$, increasing and convex in K.)

Because the preceding properties remain true when we take expectations, we see that

• $P(s, t, K, r, \sigma)$ is decreasing and convex in s.

• $P(s, t, K, r, \sigma)$ is decreasing and convex in r.

• $P(s, t, K, r, \sigma)$ is increasing and convex in K.

Moreover, because $C(s, t, K, r, \sigma)$ is increasing in σ, it follows from (7.7) that

• $P(s, t, K, r, \sigma)$ is increasing in σ.

Finally,

• $P(s, t, K, r, \sigma)$ is not necessarily increasing or decreasing in t.

The partial derivatives of $P(s, t, K, r, \sigma)$ can be obtained by using (7.6) in conjunction with the corresponding partial derivatives of $C(s, t, K, r, \sigma)$.

7.7 Exercises

Unless otherwise mentioned, the unit of time should be taken as one year.

Exercise 7.1 If the volatility of a stock is .33, find the standard deviation of

(a) $\log\left(\frac{S_d(n)}{S_d(n-1)}\right)$,

(b) $\log\left(\frac{S_m(n)}{S_m(n-1)}\right)$,

where $S_d(n)$ and $S_m(n)$ are the prices of the security at the end of day n and month n (respectively).

Exercise 7.2 The prices of a certain security follow a geometric Brownian motion with parameters $\mu = .12$ and $\sigma = .24$. If the security's price is presently 40, what is the probability that a call option, having four months until its expiration time and with a strike price of $K = 42$, will be exercised? (A security whose price at the time of expiration of a call option is above the strike price is said to finish *in the money*.)

Exercise 7.3 If the interest rate is 8%, what is the risk-neutral valuation of the call option specified in Exercise 7.2?

Exercise 7.4 What is the risk-neutral valuation of a six-month European put option to sell a security for a price of 100 when the current price is 105, the interest rate is 10%, and the volatility of the security is .30?

Exercise 7.5 A security's price follows geometric Brownian motion with drift parameter .06 and volatility parameter .3.

(a) What is the probability that the price of the security in six months is less than 90% of what it is today?

(b) Consider a newly instituted investment that, for an initial cost of A, returns you 100 in six months if the price at that time is less than 90% of what it initially was but returns you 0 otherwise. What must be the value of A in order for this investment's introduction *not* to allow an arbitrage? Assume $r = .05$.

Exercise 7.6 The price of a certain security follows a geometric Brownian motion with drift parameter $\mu = .05$ and volatility parameter $\sigma = .3$. The present price of the security is 95.

(a) If the interest rate is 4%, find the no-arbitrage cost of a call option that expires in three months and has exercise price 100.

(b) What is the probability that the call option in part (a) is worthless at the time of expiration?

(c) Suppose that a new type of investment on the security is being traded. This investment returns 50 at the end of one year if the price six months after purchasing the investment is at least 105 *and* the price one year after purchase is at least as much as the price was after six months. Determine the no-arbitrage cost of this investment.

Exercise 7.7 A European *cash-or-nothing* call pays its holder a fixed amount F if the price at expiration time is larger than K and pays 0 otherwise. Find the risk-neutral valuation of such a call – one that expires in six month's time and has $F = 100$ and $K = 40$ – if the present price of the security is 38, its volatility is .32, and the interest rate is 6%.

Exercise 7.8 If the drift parameter of the geometric Brownian motion is 0, find the expected payoff of the asset-or-nothing call in Exercise 7.7.

Exercise 7.9 To determine the probability that a European call option finishes in the money (see Exercise 7.2), is it enough to specify the five parameters K, $S(0)$, r, t, and σ? Explain your answer; if it is "no," what else is needed?

Exercise 7.10 The price of a security follows a geometric Brownian motion with drift parameter 0.05 and volatility parameter 0.4. The current price of the security is 100. A new investment that is being marketed costs 10; after 1 year the investment will pay 5 if $S(1) < 95$, will pay x if $S(1) > 110$, and will pay 0 otherwise. The nominal interest rate is 6 percent, continuously compounded.

(a) What must be the value of x if this new investment, which can be bought or sold at any level, is not to give rise to an arbitrage?
(b) What is the probability that $S(1) < 95$?

Exercise 7.11 The price of a traded security follows a geometric Brownian motion with drift 0.06 and volatility 0.4. Its current price is 40. A brokerage firm is offering, at cost C, an investment that will pay 100 at the end of 1 year either if the price of the security at 6 months is at least 42 or if the price of the security at 1 year is at least 5 percent above its price at 6 months. That is, the payoff occurs if either $S(0.5) \geq 42$ or $S(1) > 1.05\, S(0.5)$. The continuously compounded interest rate is 0.06.

(a) If this investment is not to give rise to an arbitrage, what is C?
(b) What is the probability the investment makes money for its buyer?

Exercise 7.12 The price of a traded security follows a geometric Brownian motion with drift 0.04 and volatility 0.2. Its current price is 40. A brokerage firm is offering, at cost 10, an investment that will pay 100 at the end of 1 year if $S(1) > (1 + x)40$. That is, there is a payoff of 100 if the price increases by at least $100x$ percent. Assume that the continuously compounded interest rate is 0.02, and that the new investment can be bought or sold.

(a) If this investment is not to give rise to an arbitrage, what is x?
(b) What is the probability that the investment makes money for its buyer?

Exercise 7.13 A European *asset or nothing* option that expires at time t pays its holder the asset value $S(t)$ at time t if $S(t) > K$ and pays 0 otherwise. Determine the no-arbitrage cost of such an option as a function of the parameters s, t, K, r, σ.

Exercise 7.14 What should be the cost of a call option if the strike price is equal to zero?

Exercise 7.15 What should the cost of a call option become as the exercise time becomes larger and larger? Explain your reasoning (or do the mathematics).

Exercise 7.16 What should the cost of a (K, t) call option become as the volatility becomes smaller and smaller?

Exercise 7.17 Show, by plotting the curve, that $f(r) = (ae^{-rt} - b)^+$ is, for $a > 0$, decreasing and convex in r.

Exercise 7.18 Is the function $g(r) = (a - be^{-rt})^+$ concave in r when $b > 0$? Is it convex?

REFERENCES

The Black–Scholes formula was derived in [1] by solving a stochastic differential equation. The idea of obtaining it by approximating geometric Brownian motion using multiperiod binomial models was developed in [2]. References [3], [4], and [5] are popular textbooks that deal with options, although at a higher mathematical level than the present text.

[1] Black, F., and M. Scholes (1973). "The Pricing of Options and Corporate Liabilities." *Journal of Political Economy* 81: 637–59.
[2] Cox, J., S. A. Ross, and M. Rubinstein (1979). "Option Pricing: A Simplified Approach." *Journal of Financial Economics* 7: 229–64.
[3] Cox, J., and M. Rubinstein (1985). *Options Markets*. Englewood Cliffs, NJ: Prentice-Hall.
[4] Hull, J. (1997). *Options, Futures, and Other Derivatives,* 3rd ed. Englewood Cliffs, NJ: Prentice-Hall.
[5] Luenberger, D. (1998). *Investment Science*. Oxford: Oxford University Press.

8. Additional Results on Options

8.1   Introduction

In this chapter we look at some extensions of the basic call option model. In Section 8.2 we consider European call options on dividend-paying securities under three different scenarios for how the dividend is paid. In Section 8.2.1 we suppose that the dividend for each share owned is paid continuously in time at a rate equal to a fixed fraction of the price of the security. In Sections 8.2.2 and 8.2.3 we suppose that the dividend is to be paid at a specified time, with the amount paid equal to a fixed fraction of the price of the security (Section 8.2.2) or to a fixed amount (Section 8.2.3). In Section 8.3 we show how to determine the no-arbitrage price of an American put option. In Section 8.4 we introduce a model that allows for the possibilities of jumps in the price of a security. This model supposes that the security's price changes according to a geometric Brownian motion, with the exception that at random times the price is assumed to change by a random multiplicative factor. In Section 8.4.1 we derive an exact formula for the no-arbitrage cost of a call option when the multiplicative jumps have a lognormal probability distribution. In Section 8.4.2 we suppose that the multiplicative jumps have an arbitrary probability distribution; we show that the no-arbitrage cost is always at least as large as the Black–Scholes formula when there are no jumps, and we then present an approximation for the no-arbitrage cost. In Section 8.5 we describe a variety of different techniques for estimating the volatility parameter. Section 8.6 consists of comments regarding the results obtained in this and the previous chapter.

8.2   Call Options on Dividend-Paying Securities

In this section we determine the no-arbitrage price for a European call option on a stock that pays a dividend. We consider three cases that correspond to different types of dividend payments.

8.2.1   The Dividend for Each Share of the Security Is Paid Continuously in Time at a Rate Equal to a Fixed Fraction f of the Price of the Security

For instance, if the stock's price is presently S, then in the next dt time units the dividend payment per share of stock owned will be approximately $fS\,dt$ when dt is small.

To begin, we need a model for the evolution of the price of the security over time. One way to obtain a reasonable model is to suppose that all dividends are reinvested in the purchase of additional shares of the stock. Thus, we would be continuously adding additional shares at the rate f times the number of shares we presently own. Consequently, our number of shares is growing by a continuously compounded rate f. Therefore, if we purchased a single share at time 0, then at time t we would have e^{ft} shares with a total market value of

$$M(t) = e^{ft}S(t).$$

It seems reasonable to suppose that $M(t)$ follows a geometric Brownian motion with volatility given by, say, σ. The risk-neutral probabilities on $M(t)$ are those of a geometric Brownian motion with volatility σ and drift $r - \sigma^2/2$. Consequently, for there not to be an arbitrage, all options must be priced to be fair bets under the assumption that $e^{fy}S(y)$ ($y \geq 0$) follows such a risk-neutral geometric Brownian motion.

Consider a European option to purchase the security at time t for the price K. Under the risk-neutral probabilities on $M(t)$, we have

$$\frac{S(t)}{S(0)} = \frac{e^{-ft}M(t)}{M(0)} = e^{-ft}e^W,$$

where W is a normal random variable with mean $(r - \sigma^2/2)t$ and variance $t\sigma^2$. Thus, under the risk-neutral probabilities,

$$S(t) = S(0)e^{-ft}e^W.$$

Therefore, by the arbitrage theorem, we see that if $S(0) = s$ then the

$$\text{no-arbitrage cost of } (K, t) \text{ option} = e^{-rt}E[(S(t) - K)^+]$$
$$= e^{-rt}E[(se^{-ft}e^W - K)^+]$$
$$= C(se^{-ft}, t, K, \sigma, r),$$

where $C(s, t, K, \sigma, r)$ is the Black–Scholes formula. In other words, the no-arbitrage cost of the European (K, t) call option, when the initial price is s, is exactly what its cost would be if there were no dividends but the inital price were se^{-ft}.

8.2.2 For Each Share Owned, a Single Payment of $fS(t_d)$ Is Made at Time t_d

It is usual to suppose that, at the moment the dividend is paid, the price of a share instantaneously decreases by the amount of the dividend. (If one assumes that the price never drops by at least the amount of the dividend, then buying immediately before and selling immediately after the payment of the dividend would result in an arbitrage; hence, there must be some possibility of a drop in price of at least the amount of the dividend, and the usual assumption – which is roughly in agreement with actual data – is that the price decreases by exactly the dividend paid.) Because of this downward price jump at the moment at which the dividend is paid, it is clear that we cannot model the price of the security as a geometric Brownian motion (which has no discontinuities). However, if we again suppose that the dividend payment at time t_d is used to purchase additional shares, then we can model the market value of our shares by a geometric Brownian motion. Because the price of a share immediately after the dividend is paid is $S(t_d) - fS(t_d) = (1 - f)S(t_d)$, the dividend $fS(t_d)$ from a single share can be used to purchase $f/(1 - f)$ additional shares. Hence, starting with a single share at time 0, the market value of our portfolio at time y, call it $M(y)$, is

$$M(y) = \begin{cases} S(y) & \text{if } y < t_d, \\ \frac{1}{1-f}S(y) & \text{if } y \geq t_d. \end{cases}$$

Let us take as our model that $M(y)$ ($y \geq 0$) follows a geometric Brownian motion with volatility parameter σ. The risk-neutral probabilities for this process are that of a geometric Brownian motion with volatility parameter σ and drift parameter $r - \sigma^2/2$. For $y < t_d$, $M(y) = S(y)$; thus, when $t < t_d$, the unique no-arbitrage cost of a (K, t) option on the security is just the usual Black–Scholes cost. For $t > t_d$, note that

$$\frac{S(t)}{S(0)} = (1 - f)\frac{M(t)}{M(0)}, \quad t > t_d.$$

Thus, under the risk-neutral probabilities,

$$\frac{1}{1-f}\frac{S(t)}{S(0)} = \frac{M(t)}{M(0)} = e^W, \quad t > t_d,$$

where W is a normal random variable with mean $(r - \sigma^2/2)t$ and variance $t\sigma^2$. Thus, again under the risk-neutral probabilities,

$$S(t) = (1 - f)S(0)e^W, \quad t > t_d.$$

When $t > t_d$, it follows by the arbitrage theorem that the unique no-arbitrage cost of a European (K, t) call option, when the initial price of the security is s, is exactly what its cost would be if there were no dividends but the inital price of the security were $s(1 - f)$. That is, for $t > t_d$, the

$$\text{no-arbitrage cost of } (K, t) \text{ option} = e^{-rt}E[(S(t) - K)^+]$$

$$= e^{-rt}E[(s(1 - f)e^W - K)^+]$$

$$= C(s(1 - f), t, K, \sigma, r),$$

where $C(s, t, K, \sigma, r)$ is the Black–Scholes formula.

8.2.3   For Each Share Owned, a Fixed Amount D Is to Be Paid at Time t_d

As in the previous cases, we must first determine an appropriate model for $S(y)$ ($y \geq 0$), the price evolution of the security. To begin, note that the known dividend payment D to be made to shareholders at the known time t_d necessitates that the price of the security at time $y < t_d$ must be at least $De^{-r(t_d-y)}$. This is true because, if $S(y) < De^{-r(t_d-y)}$ for some $y < t_d$, then an arbitrage can be effected by borrowing $S(y)$ at time y and using this amount to purchase the security; the security is held through time t_d and the loan is paid off immediately after the dividend is received. Consequently, we cannot model $S(y)$ ($0 \leq y \leq t_d$) as a geometric Brownian motion.

To model the price evolution up to time t_d, it is best to separate the price of the security into two parts of which one is riskless and results from the fixed payment at time t_d. That is, let

$$S^*(y) = S(y) - De^{-r(t_d-y)}, \quad y < t_d,$$

and write

$$S(y) = De^{-r(t_d - y)} + S^*(y), \quad y < t_d.$$

It is reasonable to model $S^*(y)$, $y < t_d$, as a geometric Brownian motion, with its volatility parameter denoted by σ. Because the riskless part of the price is increasing at rate r, it is intuitive that risk-neutral probabilities would result when the drift parameter of $S^*(y)$, $y < t_d$, is $r - \sigma^2/2$. To check that this assumption on the drift would result in all bets being fair, note that under it the expected present value return from purchasing the security at time 0 and then selling at time $t < t_d$ is

$$e^{-rt}E[S(t)] = e^{-rt}De^{-r(t_d - t)} + e^{-rt}E[S^*(t)]$$

$$= De^{-rt_d} + S^*(0)$$

$$= S(0).$$

Suppose now that we want to find the no-arbitrage cost of a European call option with strike price K and expiration time $t < t_d$ when the initial price of the security is s. If $K < De^{-r(t_d - t)}$, then the option will definitely be exercised (because $S(t) \geq De^{-r(t_d - t)}$). Consequently, purchasing the option in this case is equivalent to purchasing the security. By the law of one price, the cost of the option plus the present value of the strike price must therefore equal the cost of the security. That is, if $t < t_d$ and $K < De^{-r(t_d - t)}$ then the

$$\text{no-arbitrage cost of option} = s - Ke^{-rt}.$$

Suppose now that the option expires at time $t < t_d$ and its strike price K satisfies $K \geq De^{-r(t_d - t)}$. Because $S^*(y)$ is geometric Brownian motion, we can use the risk-neutral representation

$$S^*(t) = S^*(0)e^W = (s - De^{-rt_d})e^W,$$

where W is a normal random variable with mean $(r - \sigma^2/2)t$ and variance $t\sigma^2$. The arbitrage theorem yields that the

$$\text{no-arbitrage cost of option} = e^{-rt}E[(S(t) - K)^+]$$

$$= e^{-rt}E[(S^*(t) + De^{-r(t_d - t)} - K)^+]$$

$$= e^{-rt}E[((s - De^{-rt_d})e^W$$

$$- (K - De^{-r(t_d - t)}))^+]$$

$$= C(s - De^{-rt_d}, t, K - De^{-r(t_d - t)}, \sigma, r).$$

In other words, if the dividend is to be paid after the expiration date of the option, then the no-arbitrage cost of the option is given by the Black–Scholes formula for a call option on a security whose initial price is $s - De^{-rt_d}$ and whose strike price is $K - De^{-r(t_d - t)}$.

Now consider a European call option with strike price K that expires at time $t > t_d$. Suppose the initial price of the security is s. Because the price of the security will immediately drop by the dividend amount D at time t_d, we have that

$$S(t) = S^*(t), \quad t \geq t_d.$$

Hence, assuming that the volatility of the geometric Brownian motion process $S^*(y)$ remains unchanged after time t_d, we see that the risk-neutral cost of a (K, t) call option is

$$e^{-rt}E[(S(t) - K)^+] = e^{-rt}E[(S^*(t) - K)^+]$$

$$= e^{-rt}E[(S^*(0)e^W - K)^+]$$

$$= e^{-rt}E[((s - De^{-rt_d})e^W - K)^+].$$

Because the right side of the preceding equation is the Black–Scholes cost of a call option with strike price K and expiration time t, when the initial price of the security is $s - De^{-rt_d}$ we obtain that the

$$\text{risk-neutral cost of option} = C(s - De^{-rt_d}, t, K, \sigma, r).$$

In other words, if the dividend is to be paid during the life of the option, then the no-arbitrage cost of the option is given by the Black–Scholes formula – except that the initial price of the security is reduced by the present value of the dividend.

8.3   Pricing American Put Options

There is no difficulty in determining the risk-neutral prices of European put options. The put–call option parity formula gives that

$$P(s, t, K, \sigma, r) = C(s, t, K, \sigma, r) + Ke^{-rt} - s,$$

where $P(s, t, K, \sigma, r)$ is the risk-neutral price of a European put having strike price K at exercise time t, given that the price at time 0 is s, the volatility of the stock is σ, and the interest rate is r, and where

$C(s, t, K, \sigma, r)$ is the corresponding risk-neutral price for the call op-tion. However, because early exercise is sometimes beneficial, the risk-neutral pricing of American put options is not so straightforward. We will now present an efficient technique for obtaining accurate approximations of these prices.

The risk-neutral price of an American put option is the expected present value of owning the option under the assumption that the prices of the underlying security change in accordance with the risk-neutral geometric Brownian motion and that the owner utilizes an optimal policy in determining when, if ever, to exercise that option. To approximate this price, we approximate the risk-neutral geometric Brownian motion process by a multiperiod binomial process as follows. Choose a number n and, with t equal to the exercise time of the option, let $t_k = kt/n$ ($k = 0, 1, \ldots, n$). Now suppose that:

(1) the option can only be exercised at one of the times t_k ($k = 0, 1, \ldots, n$); and
(2) if $S(t_k)$ is the price of the security at time t_k, then

$$S(t_{k+1}) = \begin{cases} uS(t_k) & \text{with probability } p, \\ dS(t_k) & \text{with probability } 1 - p, \end{cases}$$

where

$$u = e^{\sigma\sqrt{t/n}}, \quad d = e^{-\sigma\sqrt{t/n}},$$

$$p = \frac{1 + rt/n - d}{u - d}.$$

The first two possible price movements of this process are indicated in Figure 8.1.

We know from Section 7.1 that the preceding discrete time approximation becomes the risk-neutral geometric Brownian motion process as n becomes larger and larger; in addition, because the price curve under geometric Brownian motion can be shown to be continuous, it is intuitive (and can be verified) that the expected loss incurred in allowing the option only to be exercised at one of the times t_k goes to 0 as n becomes larger. Hence, by choosing n reasonably large, the risk-neutral price of the American option can be accurately approximated by the expected present value return from the option, assuming that both conditions (1) and (2) hold and also that an optimal policy is employed in determining when to exercise the option. We now show how to determine this expected return.

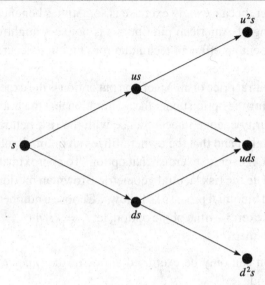

Figure 8.1: Possible Prices of the Discrete Approximation Model

To start, note that if i of the first k price movements were increases and $k - i$ were decreases, then the price at time t_k would be

$$S(t_k) = u^i d^{k-i} s.$$

Since i must be one of the values $0, 1, \ldots, k$, it follows that there are $k + 1$ possible prices of the security at time t_k. Now, let $V_k(i)$ denote the time-t_k expected return from the put, given that the put has not been exercised before time t_k, that the price at time t_k is $S(t_k) = u^i d^{k-i} s$, and that an optimal policy will be followed from time t_k onward.

To determine $V_0(0)$, the expected present value return of owning the put, we work backwards. That is, first we determine $V_n(i)$ for each of its $n + 1$ possible values of i; then we determine $V_{n-1}(i)$ for each of its n possible values of i; then $V_{n-2}(i)$ for each of its $n - 1$ possible values of i; and so on. To accomplish this task, note first that, because the option expires at time t_n,

$$V_n(i) = \max(K - u^i d^{n-i} s, 0), \tag{8.0}$$

which determines all the values $V_n(i)$, $i = 0, \ldots, n$. Now let

$$\beta = e^{-rt/n};$$

suppose we are at time t_k, the put has not yet been exercised, and the price of the stock is $u^i d^{k-i} s$. If we exercise the option at this point, then we will receive $K - u^i d^{k-i} s$. On the other hand, if we do not exercise then the price at time t_{k+1} will be either $u^{i+1} d^{k-i} s$ with probability p or $u^i d^{k-i+1} s$ with probability $1 - p$. If it is $u^{i+1} d^{k-i} s$ and we employ an optimal policy from that time on, then the time-t_k expected return from the put is $\beta V_{k+1}(i + 1)$; similarly, the expected return if the price decreases is $\beta V_{k+1}(i)$. Hence, because the price will increase with probability p or decrease with probability $1 - p$, it follows that the expected time-t_k return if we do not exercise but then continue optimally is

$$p\beta V_{k+1}(i + 1) + (1 - p)\beta V_{k+1}(i).$$

Because $K - u^i d^{k-i} s$ is the return if we exercise and because the preceding is the maximal expected return if we do not exercise, it follows that the maximal possible expected return is the larger of these two. That is, for $k = 0, \ldots, n - 1$,

$$V_k(i) = \max(K - u^i d^{k-i} s, \; \beta p V_{k+1}(i + 1) + \beta(1 - p)V_{k+1}(i)),$$
$$i = 0, \ldots, k. \quad (8.1)$$

To obtain the approximation, we first use Equation (8.0) to determine the values of $V_n(i)$; we then use Equation (8.1) with $k = n - 1$ to obtain the values $V_{n-1}(i)$; we then use Equation (8.1) with $k = n - 2$ to obtain the values $V_{n-2}(i)$; and so on until we have the desired value of $V_0(0)$, the approximation of the risk-neutral price of the American put option. Although computationally messy when done by hand, this procedure is easily programmed and can also be done with a spreadsheet.

Remarks. 1. The computations can be simplified by noting that $ud = 1$ and also by making use of the following results, which can be shown to hold.

(a) If the put is worthless at time t_k when the price of the security is x, then it is also worthless at time t_k when the price of the security is greater than x. That is,

$$V_k(i) = 0 \implies V_k(j) = 0 \text{ if } j > i.$$

(b) If it is optimal to exercise the put option at time t_k when the price is x, then it is also optimal to exercise it at time t_k when the price of the security is less than x. That is,

$$V_k(i) = K - u^i d^{k-i} s \implies V_k(j) = K - u^j d^{k-j} s \quad \text{if } j < i.$$

2. Although we defined β as $e^{-rt/n}$, we could just as well have defined it to equal $\frac{1}{1+rt/n}$.

3. The method employed to determine the values $V_k(i)$ is known as dynamic programming. We will also utilize this technique in Chapter 10, which deals with optimization models in finance.

Example 8.3a   Suppose we want to price an American put option having the following parameters:

$$s = 9, \quad t = .25, \quad K = 10, \quad \sigma = .3, \quad r = .06.$$

To illustrate the procedure, suppose we let $n = 5$ (which is much too small for an accurate approximation). With the preceding parameters, we have that

$$u = e^{.3\sqrt{.05}} = 1.0694,$$

$$d = e^{-.3\sqrt{.05}} = 0.9351,$$

$$p = 0.5056,$$

$$1 - p = 0.4944,$$

$$\beta = e^{-rt/n} = 0.997.$$

The possible prices of the security at time t_5 are:

$$9d^5 = 6.435,$$

$$9ud^4 = 7.359,$$

$$9u^2 d^3 = 8.416,$$

$$9u^3 d^2 = 9.625,$$

$$9u^i d^{5-i} > 10 \quad (i = 4, 5).$$

Hence,

$$V_5(0) = 3.565,$$

$$V_5(1) = 2.641,$$

$$V_5(2) = 1.584,$$

$$V_5(3) = 0.375,$$

$$V_5(i) = 0 \quad (i = 4, 5).$$

Since $9u^2d^2 = 9$, Equation (8.1) gives

$$V_4(2) = \max(1, \ \beta p V_5(3) + \beta(1 - p)V_5(2)) = 1,$$

which shows that it is optimal to exercise the option at time t_4 when the price is 9. From Remark 1(b) it follows that the option should also be exercised at this time at any lower price, so

$$V_4(1) = 10 - 9ud^3 = 2.130$$

and

$$V_4(0) = 10 - 9d^4 = 3.119.$$

As $9u^3d = 10.293$, Equation (8.1) gives

$$V_4(3) = \beta p V_5(4) + \beta(1 - p)V_5(3) = 0.181.$$

Similarly,

$$V_4(4) = \beta p V_5(5) + \beta(1 - p)V_5(4) = 0.$$

Continuing, we obtain

$$V_3(0) = \max(2.641, \ \beta p V_4(1) + \beta(1 - p)V_4(0)) = 2.641,$$

$$V_3(1) = \max(1.584, \ \beta p V_4(2) + \beta(1 - p)V_4(1)) = 1.584,$$

$$V_3(2) = \max(0.375, \ \beta p V_4(3) + \beta(1 - p)V_4(2)) = 0.584,$$

$$V_3(3) = \beta p V_4(4) + \beta(1 - p)V_4(3) = 0.089.$$

Similarly,

$$V_2(0) = \max(2.130, \ \beta p V_3(1) + \beta(1 - p)V_3(0)) = 2.130,$$

$$V_2(1) = \max(1, \ \beta p V_3(2) + \beta(1 - p)V_3(1)) = 1.075,$$

$$V_2(2) = \beta p V_3(3) + \beta(1 - p)V_3(2) = 0.333,$$

and

$$V_1(0) = \max(1.584, \ \beta p V_2(1) + \beta(1 - p)V_2(0)) = 1.592,$$

$$V_1(1) = \max(0.375, \ \beta p V_2(2) + \beta(1 - p)V_2(1)) = 0.698,$$

which gives the result

$$V_0(0) = \max(1, \ \beta p V_1(1) + \beta(1 - p)V_1(0)) = 1.137.$$

That is, the risk-neutral price of the put option is approximately 1.137. (The exact answer, to three decimal places, is 1.126, indicating a very respectable approximation given the small value of n that was used.) □

8.4   Adding Jumps to Geometric Brownian Motion

One of the drawbacks of using geometric Brownian motion as a model for a security's price over time is that it does not allow for the possibility of a discontinuous price jump in either the up or down direction. (Under geometric Brownian motion, the probability of having a jump would, in theory, equal 0.) Because such jumps do occur in practice, it is advantageous to consider a model for price evolution that superimposes random jumps on a geometric Brownian motion. We now consider such a model.

Let us begin by considering the times at which the jumps occur. We will suppose, for some positive constant λ, that in any time interval of length h there will be a jump with probability approximately equal to λh when h is very small. Moreover, we will assume that this probability is unchanged by any information about earlier jumps. If we let $N(t)$ denote the number of jumps that occur by time t then, under the preceding assumptions, $N(t)$, $t \geq 0$, is called a *Poisson process,* and it can be shown that

$$P\{N(t) = n\} = e^{-\lambda t}\frac{(\lambda t)^n}{n!}, \quad n = 0, 1, \dots .$$

Let us also suppose that, when the ith jump occurs, the price of the security is multiplied by the amount J_i, where $J_1, J_2, \dots$ are independent random variables having a common specified probability distribution. Further, this sequence is assumed to be independent of the times at which the jumps occur.

To complete our description of the price evolution, let $S(t)$ denote the price of the security at time t, and suppose that

$$S(t) = S^*(t) \prod_{i=1}^{N(t)} J_i, \quad t \geq 0,$$

where $S^*(t)$, $t \geq 0$, is a geometric Brownian motion, say with volatility parameter σ and drift parameter μ, that is independent of the J_i and of the times at which the jumps occur, and where $\prod_{i=1}^{N(t)} J_i$ is defined to equal 1 when $N(t) = 0$.

To find the risk-neutral probabilities for the price evolution, let

$$J(t) = \prod_{i=1}^{N(t)} J_i.$$

It will be shown in Section 8.7 that

$$E[J(t)] = e^{-\lambda t(1 - E[J])}, \tag{8.2}$$

where $E[J] = E[J_i]$ is the expected value of a multiplicative jump. Because $S^*(t)$, $t \geq 0$, is a geometric Brownian motion with parameters μ and σ, we have

$$E[S^*(t)] = S^*(0)e^{(\mu + \sigma^2/2)t}.$$

Therefore,

$$E[S(t)] = E[S^*(t)J(t)]$$

$$= E[S^*(t)]E[J(t)] \quad \text{(by independence)}$$

$$= S^*(0)e^{(\mu + \sigma^2/2 - \lambda(1 - E[J]))t}.$$

Consequently, security-buying bets will be fair bets (i.e., $E[S(t)] = S(0)e^{rt}$) provided that

$$\mu + \sigma^2/2 - \lambda(1 - E[J]) = r.$$

In other words, risk-neutral probabilities for the security's price evolution will result when μ, the drift parameter of the geometric Brownian motion $S^*(t)$, $t \geq 0$, is given by

$$\mu = r - \sigma^2/2 + \lambda - \lambda E[J].$$

By the arbitrage theorem, if all options are priced to be fair bets with respect to the preceding risk-neutral probabilities, then no arbitrage is possible. For instance, the no-arbitrage cost of a European call option having strike price K and expiration time t is given by

$$\text{no-arbitrage cost} = E[e^{-rt}(S(t) - K)^+]$$

$$= e^{-rt}E[(J(t)S^*(t) - K)^+]$$

$$= e^{-rt}E[(J(t)se^W - K)^+], \qquad (8.3)$$

where $s = S^*(0)$ is the initial price of the security and W is a normal random variable with mean $(r - \sigma^2/2 + \lambda - \lambda E[J])t$ and variance $t\sigma^2$.

In Section 8.4.1 we explicitly evaluate Equation (8.3) when the J_i are lognormal random variables, and in Section 8.4.2 we derive an approximation in the case of a general jump distribution. As always, $C(s, t, K, \sigma, r)$ will be the Black–Scholes formula.

8.4.1   *When the Jump Distribution Is Lognormal*

If the jumps J_i have a lognormal distribution with mean parameter μ_0 and variance parameter σ_0^2, then

$$E[J] = \exp\{\mu_0 + \sigma_0^2/2\}.$$

If we let

$$X_i = \log(J_i), \quad i \geq 1,$$

then the X_i are independent normal random variables with mean μ_0 and variance σ_0^2. Also,

$$J(t) = \prod_{i=1}^{N(t)} J_i = \prod_{i=1}^{N(t)} e^{X_i} = \exp\left\{\sum_{i=1}^{N(t)} X_i\right\}.$$

Consequently, using Equation (8.3), we see that the no-arbitrage cost of a European call option having strike price K and expiration time t is

$$\text{no-arbitrage cost} = e^{-rt}E\left[\left(s \exp\left\{W + \sum_{i=1}^{N(t)} X_i\right\} - K\right)^+\right], \qquad (8.4)$$

where s is the initial price of the security. Now suppose that there were a total of n jumps by time t. That is, suppose it were known that $N(t) = n$.

Then $W + \sum_{i=1}^{N(t)} X_i$ would be a normal random variable with mean and variance given by

$$E\left[W + \sum_{i=1}^{N(t)} X_i \mid N(t) = n\right] = (r - \sigma^2/2 + \lambda - \lambda E[J])t + n\mu_0,$$

$$\text{Var}\left(W + \sum_{i=1}^{N(t)} X_i \mid N(t) = n\right) = t\sigma^2 + n\sigma_0^2.$$

Therefore, if we let

$$\sigma^2(n) = \sigma^2 + n\sigma_0^2/t$$

and let

$$r(n) = r - \sigma^2/2 + \lambda - \lambda E[J] + \frac{n\mu_0}{t} + \sigma^2(n)/2$$

$$= r + \lambda - \lambda E[J] + \frac{n}{t}(\mu_0 + \sigma_0^2/2)$$

$$= r + \lambda - \lambda E[J] + \frac{n}{t} \log(E[J]), \tag{8.5}$$

then it follows, when $N(t) = n$, that $W + \sum_{i=1}^{N(t)} X_i$ is a normal random variable with variance $t\sigma^2(n)$ and mean $(r(n) - \sigma^2(n)/2)t$. But this implies that, when $N(t) = n$,

$$e^{-r(n)t}E\left[\left(s \exp\left\{W + \sum_{i=1}^{N(t)} X_i\right\} - K\right)^+ \mid N(t) = n\right]$$

$$= C(s, t, K, \sigma(n), r(n)).$$

Multiplying both sides of the preceding equation by $e^{(r(n)-r)t}$ gives

$$e^{-rt}E\left[\left(s \exp\left\{W + \sum_{i=1}^{N(t)} X_i\right\} - K\right)^+ \mid N(t) = n\right]$$

$$= e^{(r(n)-r)t}C(s, t, K, \sigma(n), r(n)).$$

Equation (8.4) shows that the preceding expression is the desired expected value if we are given that there are n jumps by time t. Consequently, it is reasonable (and can be shown to be correct) that the unconditional expected value should be a weighted average of these quantities, with the weight given to the quantity indexed by n equal to the probability that $N(t) = n$. That is,

no-arbitrage cost

$$= \sum_{n=0}^{\infty} e^{-\lambda t} \frac{(\lambda t)^n}{n!} e^{(r(n)-r)t} C(s, t, K, \sigma(n), r(n))$$

$$= \sum_{n=0}^{\infty} e^{-\lambda t E[J]} (E[J])^n \frac{(\lambda t)^n}{n!} C(s, t, K, \sigma(n), r(n)) \qquad \text{(from (8.5))}$$

$$= \sum_{n=0}^{\infty} e^{-\lambda t E[J]} \frac{(\lambda t E[J])^n}{n!} C(s, t, K, \sigma(n), r(n)).$$

Summing up, we have proved the following.

Theorem 8.4.1 *If the jumps have a lognormal distribution with mean parameter μ_0 and variance parameter σ_0^2, then the no-arbitrage cost of a European call option having strike price K and expiration time t is as follows:*

$$\textit{no-arbitrage cost} = \sum_{n=0}^{\infty} e^{-\lambda t E[J]} \frac{(\lambda t E[J])^n}{n!} C(s, t, K, \sigma(n), r(n)),$$

where

$$\sigma^2(n) = \sigma^2 + n\sigma_0^2/t,$$

$$r(n) = r + \lambda(1 - E[J]) + \frac{n}{t} \log(E[J]),$$

and

$$E[J] = \exp\{\mu_0 + \sigma_0^2/2\}.$$

Remark. Although Theorem 8.4.1 involves an infinite series, in most applications λ – the rate at which jumps occur – will be quite small and thus the sum will converge rapidly.

8.4.2   *When the Jump Distribution Is General*

We start with Equation (8.3), which states that the no-arbitrage cost of a European call option having strike price K and expiration time t is as follows:

$$\text{no-arbitrage cost} = e^{-rt} E[(J(t)se^W - K)^+],$$

where s is the price of the security at time 0 and W is a normal random variable with mean $(r - \sigma^2/2 + \lambda - \lambda E[J])t$ and variance $t\sigma^2$. If we let

$$W^* = W - \lambda t(1 - E[J])$$

and

$$s_t = s e^{\lambda t(1 - E[J])} = \frac{s}{E[J(t)]},$$

then we can write

$$\text{no-arbitrage cost} = E[e^{-rt}(s_t J(t) e^{W^*} - K)^+].$$

Because W^* is a normal random variable with mean $(r - \sigma^2/2)t$ and variance $t\sigma^2$, it follows that

$$\text{no-arbitrage cost} = E[C(s_t J(t), t, K, \sigma, r)]. \tag{8.6}$$

Because $C(s, t, K, \sigma, r)$ is a convex function of s, it follows from a result known as Jensen's inequality (see Section 9.2) that

$$E[C(s_t J(t), t, K, \sigma, r)] \geq C(E[s_t J(t)], t, K, \sigma, r) = C(s, t, K, \sigma, r),$$

thus showing that the no-arbitrage cost in the jump model is not less than it is in the same model excluding jumps. (Actually, it will be strictly larger in the jump model provided that $P\{J_i = 1\} \neq 1$.)

An approximation for the no-arbitrage cost can be obtained by regarding $C(x) = C(x, t, K, \sigma, r)$ solely as a function of x (by keeping the other variables fixed), expanding it in a Taylor series about some value x_0, and then ignoring all terms beyond the third to obtain

$$C(x) \approx C(x_0) + C'(x_0)(x - x_0) + C''(x_0)(x - x_0)^2/2.$$

Therefore, for any nonnegative random variable X, we have

$$C(X) \approx C(x_0) + C'(x_0)(X - x_0) + C''(x_0)(X - x_0)^2/2.$$

Letting $x_0 = E[X]$ and taking expectations of both sides of the preceding yields that

$$E[C(X)] \approx C(E[X]) + C''(E[X]) \operatorname{Var}(X)/2.$$

Therefore, letting

$$X = s_t J(t), \qquad E[X] = s$$

gives that

$$E[C(s_t J(t))] \approx C(s) + C''(s)s_t^2 \operatorname{Var}(J(t))/2.$$

It can now be shown (see Section 8.7) that

$$\operatorname{Var}(J(t)) = e^{-\lambda t(1 - E[J^2])} - e^{-2\lambda t(1 - E[J])}, \qquad (8.7)$$

where J has the probability distribution of the J_i. Therefore, using the formula derived in Section 7.5 for $C''(s)$ (which is called gamma in that section) leads to the approximation given in the following theorem, which sums up the results of this subsection.

Theorem 8.4.2 *Assuming a general distribution for the size of a jump, the*

$$no\text{-}arbitrage\ option\ cost = E[C(s_t J(t), t, K, \sigma, r)]$$

$$\geq C(s, t, K, \sigma, r).$$

Moreover,

no-arbitrage option cost

$$\approx C(s, t, K, \sigma, r) + s_t^2 [e^{-\lambda t(1 - E[J^2])} - e^{-2\lambda t(1 - E[J])}] \frac{1}{2s\sigma\sqrt{2\pi t}} e^{-\omega^2/2}$$

$$= C(s, t, K, \sigma, r) + s^2 (e^{\lambda t(1 - 2E[J] + E[J^2])} - 1) \frac{1}{2s\sigma\sqrt{2\pi t}} e^{-\omega^2/2},$$

where

$$s_t = se^{\lambda t(1 - E[J])}$$

and

$$\omega = \frac{rt + \sigma^2 t/2 - \log(K/s)}{\sigma\sqrt{t}}.$$

8.5   Estimating the Volatility Parameter

Whereas four of the five parameters needed to evaluate the Black–Scholes formula – namely, s, t, K, and r – are known quantities, the value of σ has to be estimated. One approach is to use historical data. Section 8.5.1 gives the standard approach for estimating a population variance; Section 8.5.2 applies the standard approach to obtain an estimator of σ based on closing prices of the security over successive days; Section 8.5.3 gives an improved estimator based on both daily closing and opening prices; and Section 8.5.4 gives a more sophisticated estimator that uses daily high and low prices as well as daily opening and closing prices.

8.5.1 Estimating a Population Mean and Variance

Suppose that $X_1, \ldots, X_n$ are independent random variables having a common probability distribution with mean μ_0 and variance σ_0^2. The average of these data values,

$$\bar{X} = \frac{\sum_{i=1}^{n} X_i}{n},$$

is the usual estimator of the mean. Because

$$\sigma_0^2 = \text{Var}(X_i) = E[(X_i - \mu_0)^2],$$

it would appear that σ_0^2 could be estimated by

$$\frac{\sum_{i=1}^{n} (X_i - \mu_0)^2}{n}.$$

However, this estimator cannot be directly utilized when the mean μ_0 is unknown. To use it, we must first replace the unknown μ_0 by its estimator $\bar{X}$. If we then replace n by $n - 1$, we obtain the *sample variance* S^2, defined by

$$S^2 = \frac{\sum_{i=1}^{n} (X_i - \bar{X})^2}{n - 1}.$$

The sample variance is the standard estimator of the variance σ_0^2. It is an *unbiased* estimator of σ_0^2, meaning that

$$E[S^2] = \sigma_0^2.$$

(It is because we wanted the estimator to be unbiased that we changed its denominator from n to $n - 1$.) The effectiveness of S^2 as an estimator of the variance can be measured by its mean square error (MSE), defined as

$$\text{MSE} = E[(S^2 - \sigma_0^2)^2]$$
$$= \text{Var}(S^2).$$

When the X_i come from a normal distribution, it can be shown that

$$\text{Var}(S^2) = \frac{2\sigma_0^4}{n-1}. \tag{8.8}$$

8.5.2   The Standard Estimator of Volatility

Suppose that we want to estimate σ using t time units of historical data, which we will suppose run from time 0 to time t. That is, suppose that the present time is t and that we have the historical price data $S(y)$, $0 \leq y \leq t$. Fix a positive integer n, let $\ell = t/n$, and define the random variables

$$X_1 = \log\left(\frac{S(\ell)}{S(0)}\right),$$

$$X_2 = \log\left(\frac{S(2\ell)}{S(\ell)}\right),$$

$$X_3 = \log\left(\frac{S(3\ell)}{S(2\ell)}\right),$$

$$\vdots$$

$$X_n = \log\left(\frac{S(n\ell)}{S((n-1)\ell)}\right).$$

Under the assumption that the price evolution follows a geometric Brownian motion with parameters μ and σ, it follows that $X_1, \ldots, X_n$ are independent normal random variables with mean $\ell\mu$ and variance $\ell\sigma^2$. From Section 8.5.1, it follows that we can use $\sum_{i=1}^{n}(X_i - \bar{X})^2/(n-1)$ to estimate $\ell\sigma^2$. Therefore, we can estimate σ^2 by

$$\widehat{\sigma^2} = \frac{1}{\ell}\frac{\sum_{i=1}^{n}(X_i - \bar{X})^2}{n-1}.$$

Moreover, it follows from Equation (8.8) that

$$\text{Var}(\widehat{\sigma^2}) = \frac{1}{\ell^2}\frac{2(\ell\sigma^2)^2}{n-1} = \frac{2\sigma^4}{n-1}. \tag{8.9}$$

It follows from Equation (8.9) that we can use price data history over any time interval to obtain an arbitrarily precise estimator of σ^2. That is, breaking up the time interval into a large number of subintervals results in an unbiased estimator of σ^2 having an arbitrarily small variance. The difficulty with this approach, however, is that it strongly depends on the assumption that the logarithms of price ratios $S(i\ell)/S((i-1)\ell)$ are independent with a common distribution, even when the time lag ℓ is arbitrarily small. Indeed, even assuming that a security's price history resembles a geometric Brownian motion process, it is unlikely to look like one under a microscope. That is, while successive daily closing prices might appear to be consistent with a geometric Brownian motion, it is unlikely that this would be true for hourly (or more frequent) prices. For this reason we recommend that the preceding procedure be used with ℓ equal to one day. Because the unit of time is one year and there are approximately 252 trading days in a year, $\ell = 1/252$.

To use this method to estimate σ, consider n successive daily closing prices $C_1, \ldots, C_n$, where C_i is the closing price on trading day i. Let C_0 be the closing price of the security immediately before these n days, and set

$$X_i = \log\left(\frac{C_i}{C_{i-1}}\right) = \log(C_i) - \log(C_{i-1}).$$

The sample variance of these data values,

$$S^2 = \frac{\sum_{i=1}^n (X_i - \bar{X})^2}{n-1},$$

can be taken as the estimator of $\sigma^2/252$; $S\sqrt{252}$ can be used to estimate σ.

Remark. If μ and σ are the drift and volatility parameters of the geometric Brownian motion, then

$$E\left[\log\left(\frac{C_i}{C_{i-1}}\right)\right] = \frac{\mu}{252}, \qquad \sqrt{\mathrm{Var}\left(\log\left(\frac{C_i}{C_{i-1}}\right)\right)} = \frac{\sigma}{\sqrt{252}}.$$

Because μ will typically have a value close to 0 whereas σ is typically greater than .2, it follows that the mean of $X_i = \log(C_i/C_{i-1})$ is negligible with respect to its standard deviation. Therefore, we could approximate μ by 0 and, with very small loss of efficiency, use

$$\frac{\sum_{i=1}^n X_i^2}{n}$$

as the estimator of $\sigma^2/252$. It is important to note that this estimator can be used even when the geometric Brownian motion has a time-varying drift parameter. (Recall that the Black–Scholes formula yields the unique no-arbitrage cost even in the case of a time-varying drift parameter.)

8.5.3   Using Opening and Closing Data

Let C_i denote the (closing) price of a security at the end of trading day i. Under the assumption that the security's price follows a geometric Brownian motion, $\log(C_i/C_{i-1})$ is a normal random variable whose mean is approximately 0 and whose variance is $\sigma^2/252$. Letting O_i be the opening price of the security at the beginning of trading day i, we can write

$$\log\left(\frac{C_i}{C_{i-1}}\right) = \log\left(\frac{C_i}{O_i}\frac{O_i}{C_{i-1}}\right)$$

$$= \log\left(\frac{C_i}{O_i}\right) + \log\left(\frac{O_i}{C_{i-1}}\right).$$

Assuming that C_i/O_i and O_i/C_{i-1} are independent – that is, assuming that the ratio price change during a trading day is independent of the ratio price change that occurred while the market was closed – it follows that

$$\mathrm{Var}(\log(C_i/C_{i-1})) = \mathrm{Var}(\log(C_i/O_i)) + \mathrm{Var}(\log(O_i/C_{i-1}))$$

$$= \mathrm{Var}(C_i^* - O_i^*) + \mathrm{Var}(O_i^* - C_{i-1}^*), \quad (8.10)$$

where

$$C_j^* = \log(C_j), \qquad O_j^* = \log(O_j).$$

Because $C_i^* - O_i^*$ and $O_i^* - C_{i-1}^*$ both have a mean of approximately 0, we can estimate $\sigma^2/252 = \mathrm{Var}(\log(C_i/C_{i-1}))$ by

$$\frac{\sum_{i=1}^{n}(C_i^* - O_i^*)^2}{n} + \frac{\sum_{i=1}^{n}(O_i^* - C_{i-1}^*)^2}{n}.$$

This yields the estimator $\hat{\sigma}$ of the volatility parameter σ:

$$\hat{\sigma} = \sqrt{\frac{252}{n}\sum_{i=1}^{n}[(C_i^* - O_i^*)^2 + (O_i^* - C_{i-1}^*)^2]}. \quad (8.11)$$

Equation (8.11) should be a better estimator of σ than is the standard estimator described in Section 8.5.2.

8.5.4   *Using Opening, Closing, and High–Low Data*

Following the notation introduced in Section 8.5.3, let $X^* = \log(X)$ for any value X.

Let $H(t)$ be the highest price and $L(t)$ the lowest price of a security over an interval of length t. That is,

$$H(t) = \max_{0 \le y \le t} S(y),$$

$$L(t) = \min_{0 \le y \le t} S(y).$$

Assuming that the security's price follows geometric Brownian motion with drift 0 and volatility σ, it can be shown that

$$E[(H^*(t) - L^*(t))^2] = 2.773 \, \mathrm{Var}\left(\log\left(\frac{S(t)}{S(0)} \right) \right).$$

Now let O_i and C_i be the opening and closing prices on trading day i, and let H_i and L_i be the high and the low prices during that day. Because $E[\log(C_i/O_i)] \approx 0$, we can approximate the price history during a trading day as a geometric Brownian motion process with drift parameter 0. Therefore, using the preceding identity, we see that

$$E[(H_i^* - L_i^*)^2] \approx 2.773 \, \mathrm{Var}(\log(C_i/O_i)).$$

Thus, using n days' worth of data, we can estimate $\mathrm{Var}(\log(C_i/O_i))$ by the estimator

$$\mathcal{E}_1 = \frac{1}{2.773} \frac{\sum_{i=1}^{n}(H_i^* - L_i^*)^2}{n}$$

$$= \frac{.361}{n} \sum_{i=1}^{n}(H_i^* - L_i^*)^2.$$

However, $\mathrm{Var}(\log(C_i/O_i)) = \mathrm{Var}(C_i^* - O_i^*)$ can also be estimated by

$$\mathcal{E}_2 = \frac{1}{n} \sum_{i=1}^{n}(C_i^* - O_i^*)^2.$$

Any linear combination of these estimators of the form

$$\alpha \mathcal{E}_1 + (1 - \alpha)\mathcal{E}_2$$

can also be used to estimate $\mathrm{Var}(\log(C_i/O_i))$. The best estimator of this type (i.e., the one whose variance is smallest) can be shown to result when $\alpha = .5/.361 = 1.39$. That is, the best estimator of $\mathrm{Var}(\log(C_i/O_i))$ is

$$\mathcal{E} = \frac{.5}{.361}\mathcal{E}_1 - .39\mathcal{E}_2$$

$$= \frac{1}{n}\sum_{i=1}^{n}[.5(H_i^* - L_i^*)^2 - .39(C_i^* - O_i^*)^2]. \tag{8.12}$$

Because we can estimate $\mathrm{Var}(\log(O_i/C_{i-1})) = \mathrm{Var}(O_i^* - C_{i-1}^*)$ by $\frac{1}{n}\sum_{i=1}^{n}(O_i^* - C_{i-1}^*)^2$, it follows that

$$\mathcal{E} + \frac{1}{n}\sum_{i=1}^{n}(O_i^* - C_{i-1}^*)^2$$

$$= \frac{1}{n}\sum_{i=1}^{n}[.5(H_i^* - L_i^*)^2 - .39(C_i^* - O_i^*)^2 + (O_i^* - C_{i-1}^*)^2]$$

is an estimator of

$$\mathrm{Var}(\log(C_i/O_i)) + \mathrm{Var}(\log(O_i/C_{i-1})) = \mathrm{Var}(\log(C_i/C_{i-1}))$$

$$= \sigma^2/252.$$

Consequently, we can estimate the volatility parameter σ by

$$\hat{\sigma} = \sqrt{\frac{252}{n}\sum_{i=1}^{n}[.5(H_i^* - L_i^*)^2 - .39(C_i^* - O_i^*)^2 + (O_i^* - C_{i-1}^*)^2]}.$$

$$\tag{8.13}$$

Remark. The estimator of σ given in Equation (8.13) has not previously appeared in the literature. The approach presented here built on the work of Garman and Klass (see reference [2]), who derived the estimator of $\mathrm{Var}(\log(C_i/O_i))$ given by Equation (8.12). In their further analysis, however, Garman and Klass assume not only that the security's price follows a geometric Brownian motion when the market is open but also that it follows *the same* (although now unobservable) geometric Brownian motion while the market is closed. Based on this assumption, they supposed that

$$\text{Var}(C_i^* - O_i^*) = \frac{1-f}{252}\sigma^2,$$

$$\text{Var}(O_i^* - C_{i-1}^*) = \frac{f}{252}\sigma^2,$$

where f is the fraction of the day that the market is closed. However, this assumption – that the security's price when the market is closed changes according to the same probability law as when it is open – seems quite doubtful. Therefore, we have chosen to make the much weaker assumption that the ratio price changes O_i/C_{i-1} are independent of all prices up to market closure on day $i - 1$.

8.6   Some Comments

8.6.1   *When the Option Cost Differs from the Black–Scholes Formula*

Suppose now that we have estimated the value of σ and inserted that value into the Black–Scholes formula to obtain $C(s, t, K, \sigma, r)$. What if the market price of the option is unequal to $C(s, t, K, \sigma, r)$? Practically speaking, is there really a strategy that yields us a sure win?

Unfortunately, the answer to this question is "probably not." For one thing, the arbitrage strategy when the actual trading price for the option differs from that given by the Black–Scholes formula requires that one continuously trade (buy or sell) the underlying security. Not only is this physically impossible, but even if discretely approximated it might (in practice) result in large transaction costs that could easily exceed the gain of the arbitrage. A second reason for our answer is that even if we are willing to accept that our estimate of the historical value of σ is very precise, it is possible that its value might change over the option's life. Indeed, perhaps one reason that the market price differs from the formula is because "the market" believes that the stock's volatility over the life of the option will not be the same as it was historically. Indeed, it has been suggested that – rather than using historical data to estimate a security's volatility – a more accurate estimate can often be obtained by finding the value of σ that, along with the other parameters

$(s, t, K,$ and $r)$ of the option, makes the Black–Scholes valuation equal to the actual market cost of the option. However, one difficulty with this *implied volatility* is that different options on the same security, having either different expiration times or strike prices or both, will often give rise to different implied volatility estimates of σ. A common occurrence is that implied volatilities derived from far out-of-the-money call options (i.e., ones in which the present market price is far below the strike price) are larger than ones derived from at-the-money options (where the present price is near the strike price). With respect to the Black–Scholes valuation based on estimating σ via historical data, these comments suggest that out-of-the-money call options tend to be overpriced with respect to at-the-money call options. A third (even more basic) reason why there is probably no way to guarantee a win is that the assumption that the underlying security follows a geometric Brownian motion is only an approximation to reality, and – even ignoring transaction costs – the existence of an arbitrage strategy relies on this assumption. Indeed many traders would argue against the geometric Brownian motion assumption that future price changes are independent of past prices, claiming to the contrary that past prices are often an indication of an upward or downward trend in future prices.

8.6.2   *When the Interest Rate Changes*

We have previously shown that the option cost is an increasing function of the interest rate. Does this imply that the cost of an option should increase if the central bank announces an increase in the interest rate (say, on U.S. treasuries) and should decrease if the bank anounces a decrease in the interest rate? The answer is yes, *provided* that the security's volatility remains the same. However, one should be careful about making the assumption that a security's volatility will remain unchanged when there is a change in interest rates. An increase in interest rates often has the effect of causing some investors to switch from stocks to either bonds or investments having a fixed return rate, with the reverse resulting when there is a decrease in interest rates; such actions will probably result in a change in the volatility of a security.

8.6.3   *Final Comments*

If you believe that geometric Brownian motion is a reasonable (albeit approximate) model, then the Black–Scholes formula gives a reasonable option price. If this price is significantly above (below) the market price, then a strategy involving buying (selling) options and selling (buying) the underlying security can be devised. Such a strategy, although not yielding a certain win, can often yield a gain that has a positive expected value along with a small variance.

Under the assumption that the security's price over time follows a geometric Brownian motion with parameters μ and σ, one can often devise strategies that have positive expected gains and relatively small risks *even when the cost of the option is as given by the Black–Scholes formula.* For suppose that, based on an estimation using empirical data, you believe that the parameter μ is unequal to the risk-neutral value $r - \sigma^2/2$. If

$$\mu > r - \sigma^2/2$$

then both buying the security and buying the call option will result in positive expected present value gains. Although you cannot avoid all risks (since no arbitrage is possible), a low-risk strategy with a positive expected gain can be effected either by (a) introducing a risk-averse utility function and then finding a strategy that maximizes the expected utility or (b) finding a strategy that has a reasonably large expected gain along with a reasonably small variance. Such strategies would either buy some security shares and sell some calls, or the reverse. Similarly, if

$$\mu < r - \sigma^2/2$$

then both buying the security and buying the call option have negative expected present value gains, and again we can search for a low-risk, positive expectation strategy that sells one and buys the other. These types of problems are considered in the following chapter, which also introduces utility functions and their uses.

It is our opinion that the geometric Brownian motion model of the prices of a security over time can often be substantially improved upon, and that – rather than blindly assuming such a model – one could sometimes do better by using historical data to fit a more general model. If successful, the improved model can give more accurate option prices, resulting in more efficient strategies. The final two chapters of this book deal with these more general models. In Chapter 12 we show that geometric Brownian motion is not consistent with actual data on crude oil prices; an improved model is presented that allows tomorrow's closing price to depend not only on today's closing price but also on yesterday's, and a risk-neutral option price valuation based on this model is indicated. In Chapter 13 we show that a generalization of the geometric Brownian motion model results in an autoregressive model that can be used when modeling a security whose prices have a mean reverting quality.

8.7   Appendix

For the model of Section 8.4, we need to derive $E[J^m(t)]$ for $m = 1, 2$. Observe that

$$J^m(t) = \prod_{i=1}^{N(t)} J_i^m.$$

Consequently, given that $N(t) = n$, we have

$$E[J^m(t) \mid N(t) = n]$$

$$= E\left[\prod_{i=1}^{N(t)} J_i^m \mid N(t) = n\right]$$

$$= E\left[\prod_{i=1}^{n} J_i^m \mid N(t) = n\right]$$

$$= E\left[\prod_{i=1}^{n} J_i^m\right] \quad \text{(by the independence of the } J_i \text{ and } N(t)\text{)}$$

$$= (E[J^m])^n \quad \text{(by the independence of the } J_i\text{)}.$$

Therefore,

$$E[J^m(t)] = \sum_{n=0}^{\infty} E[J^m(t) \mid N(t) = n]P\{N(t) = n\}$$

$$= \sum_{n=0}^{\infty} (E[J^m])^n e^{-\lambda t}(\lambda t)^n/n!$$

$$= \sum_{n=0}^{\infty} e^{-\lambda t}(\lambda t E[J^m])^n/n!$$

$$= e^{-\lambda t(1 - E[J^m])}.$$

As a result,

$$E[J(t)] = e^{-\lambda t(1 - E[J])}$$

and

$$\text{Var}(J(t)) = E[J^2(t)] - (E[J(t)])^2 = e^{-\lambda t(1 - E[J^2])} - e^{-2\lambda t(1 - E[J])}.$$

8.8   Exercises

Exercise 8.1   Does the put–call option parity formula for European call and put options remain valid when the security pays dividends?

Exercise 8.2   For the model of Section 8.2.1, under the risk-neutral probabilities, what process does the security's price over time follow?

Exercise 8.3   Find the no-arbitrage cost of a European (K, t) call option on a security that, at times t_{d_i} $(i = 1, 2)$, pays $fS(t_{d_i})$ as dividends, where $t_{d_1} < t_{d_2} < t$.

Exercise 8.4   Consider an American (K, t) call option on a security that pays a dividend at time t_d, where $t_d < t$. Argue that the call is exercised either immediately before time t_d or at the expiration time t.

Exercise 8.5   Consider a European (K, t) call option whose return at expiration time is capped by the amount B. That is, the payoff at t is

$$\min((S(t) - K)^+, B).$$

Explain how you can use the Black–Scholes formula to find the no-arbitrage cost of this option.

Hint: Express the payoff in terms of the payoffs from two plain (uncapped) European call options.

Exercise 8.6   The current price of a security is s. Consider an investment whose cost is s and whose payoff at time 1 is, for a specified choice of β satisfying $0 < \beta < e^r - 1$, given by

$$\text{return} = \begin{cases} (1 + \beta)s & \text{if } S(1) \le (1 + \beta)s, \\ (1 + \beta)s + \alpha(S(1) - (1 + \beta)s) & \text{if } S(1) \ge (1 + \beta)s. \end{cases}$$

Determine the value of α if this investment (whose payoff is both un-capped and always greater than the initial cost of the investment) is not to give rise to an arbitrage.

Exercise 8.7 The following investment is being offered on a security whose current price is s. For an initial cost of s and for the value β of your choice (provided that $0 < \beta < e^r - 1$), your return after one year is given by

$$\text{return} = \begin{cases} (1+\beta)s & \text{if } S(1) \leq (1+\beta)s, \\ S(1) & \text{if } (1+\beta)s \leq S(1) \leq K, \\ K & \text{if } S(1) > K, \end{cases}$$

where $S(1)$ is the price of the security at the end of one year. In other words, at the price of capping your maximum return at time 1 you are guaranteed that your return at time 1 is at least $1 + \beta$ times your original payment. Show that this investment (which can be bought or sold) does not give rise to an arbitrage when K is such that

$$C(s, 1, K, \sigma, r) = C(s, 1, s(1 + \beta), \sigma, r) + s(1 + \beta)e^{-r} - s,$$

where $C(s, t, K, \sigma, r)$ is the Black–Scholes formula.

Exercise 8.8 Show that, for $f < r$,

$$C(se^{-ft}, t, K, \sigma, r) = e^{-ft}C(s, t, K, \sigma, r - f).$$

Exercise 8.9 An option on an option, sometimes called a *compound option,* is specified by the parameter pairs (K_1, t_1) and (K, t), where $t_1 < t$. The holder of such a compound option has the right to purchase, for the amount K_1, a (K, t) call option on a specified security. This option to purchase the (K, t) call option can be exercised any time up to time t_1.

(a) Argue that the option to purchase the (K, t) call option would never be exercised before its expiration time t_1.
(b) Argue that the option to purchase the (K, t) call option should be exercised if and only if $S(t_1) \geq x$, where x is the solution of

$$K_1 = C(x, t - t_1, K, \sigma, r),$$

$C(s, t, K, \sigma, r)$ is the Black–Scholes formula, and $S(t_1)$ is the price of the security at time t_1.

(c) Argue that there is a unique value x that satisfies the preceding identity.

(d) Argue that the unique no-arbitrage cost of this compound option can be expressed as

no-arbitrage cost of compound option

$$= e^{-rt_1} E[C(se^W, t - t_1, K, \sigma, r)I(se^W > x)],$$

where: $s = S(0)$ is the initial price of the security; x is the value specified in part (b); W is a normal random variable with mean $(r - \sigma^2/2)t_1$ and variance $\sigma^2 t_1$; $I(se^W > x)$ is defined to equal 1 if $se^W > x$ and to equal 0 otherwise; and $C(s, t, K, \sigma, r)$ is the Black–Scholes formula. (The no-arbitrage cost can be simplified to an expression involving bivariate normal probabilities.)

Exercise 8.10 A (K_1, t_1, K_2, t_2) *double call option* is one that can be exercised either at time t_1 with strike price K_1 or at time t_2 $(t_2 > t_1)$ with strike price K_2.

(a) Argue that you would never exercise at time t_1 if $K_1 > e^{-r(t_2-t_1)}K_2$.

(b) Assume that $K_1 < e^{-r(t_2-t_1)}K_2$. Argue that there is a value x such that the option should be exercised at time t_1 if $S(t_1) > x$ and not exercised if $S(t_1) < x$.

Exercise 8.11 Continue Figure 8.1 so that it gives the possible price patterns for times t_0, t_1, t_2, t_3, t_4.

Exercise 8.12 Using the notation of Section 8.3, which of the following statements do you think are true? Explain your reasoning.

(a) $V_k(i)$ is nondecreasing in k for fixed i.

(b) $V_k(i)$ is nonincreasing in k for fixed i.

(c) $V_k(i)$ is nondecreasing in i for fixed k.

(d) $V_k(i)$ is nonincreasing in i for fixed k.

Exercise 8.13 Give the risk-neutral price of a European put option whose parameters are as given in Example 8.3a.

Exercise 8.14 Derive an approximation to the risk-neutral price of an American put option having parameters

$$s = 10, \quad t = .25, \quad K = 10, \quad \sigma = .3, \quad r = .06.$$

Exercise 8.15 An American asset-or-nothing call option (with parameters K, F and expiration time t) can be exercised any time up to t. If the security's price when the option is exercised is K or higher, then the amount F is returned; if the security's price when the option is exercised is less than K, then nothing is returned. Explain how you can use the multiperiod binomial model to approximate the risk-neutral price of an American asset-or-nothing call option.

Exercise 8.16 Derive an approximation to the risk-neutral price of an American asset-or-nothing call option when

$$s = 10, \quad t = .25, \quad K = 11, \quad F = 20, \quad \sigma = .3, \quad r = .06.$$

Exercise 8.17 Table 8.1 (pp. 150–151) presents data concerning the stock prices of Microsoft from August 13 to November 1, 2001.

(a) Use this table and the estimator of Section 8.5.2 to estimate σ.
(b) Use the estimator of Section 8.5.3 to estimate σ.
(c) Use the estimator of Section 8.5.4 to estimate σ.

REFERENCES

[1] Cox, J., and M. Rubinstein (1985). *Options Markets.* Englewood Cliffs, NJ: Prentice-Hall.
[2] Garman, M., and M. J. Klass (1980). "On the Estimation of Security Price Volatilities from Historical Data." *Journal of Business* 53: 67–78.
[3] Merton, R. C. (1976). "Option Pricing When Underlying Stock Returns Are Discontinuous." *Journal of Financial Economics* 3: 125–44.
[4] Rogers, L. C. G., and S. E. Satchell (1991). "Estimating Variance from High, Low, and Closing Prices." *Annals of Applied Probability* 1: 504–12.

Table 8.1

Date	Open	High	Low	Close	Volume
12-Nov-01	64.7	66.44	63.65	65.79	28,876,400
09-Nov-01	64.34	65.65	63.91	65.21	24,006,800
08-Nov-01	64.46	66.06	63.66	64.42	37,113,900
07-Nov-01	64.22	65.05	64.03	64.25	29,449,500
06-Nov-01	62.7	64.94	62.16	64.78	34,306,000
05-Nov-01	61.86	64.03	61.75	63.27	33,200,800
02-Nov-01	61.93	63.02	60.51	61.4	41,680,000
01-Nov-01	60.08	62.25	59.6	61.84	54,835,600
31-Oct-01	59.3	60.73	58.1	58.15	32,350,000
30-Oct-01	58.92	59.54	58.19	58.88	28,697,800
29-Oct-01	62.1	62.2	59.54	59.64	27,564,700
26-Oct-01	62.32	63.63	62.08	62.2	32,254,700
25-Oct-01	60.61	62.6	59.57	62.56	37,659,100
24-Oct-01	60.5	61.62	59.62	61.32	39,570,700
23-Oct-01	60.47	61.44	59.4	60.43	40,162,500
22-Oct-01	57.9	60.18	57.47	60.16	36,161,800
19-Oct-01	57.4	58.01	55.63	57.9	45,609,800
18-Oct-01	56.34	57.58	55.5	56.75	39,174,000
17-Oct-01	59.12	59.3	55.98	56.03	36,855,300
16-Oct-01	57.87	58.91	57.21	58.45	33,084,500
15-Oct-01	55.9	58.5	55.85	58.06	34,218,500
12-Oct-01	55.7	56.64	54.55	56.38	31,653,500
11-Oct-01	55.76	56.84	54.59	56.32	41,871,300
10-Oct-01	53.6	55.75	53.0	55.51	43,174,600
09-Oct-01	57.5	57.57	54.19	54.56	49,738,800
08-Oct-01	56.8	58.65	56.74	58.04	30,302,900
05-Oct-01	56.16	58.0	54.94	57.72	40,422,200
04-Oct-01	56.92	58.4	56.21	56.44	50,889,000
03-Oct-01	52.48	56.93	52.4	56.23	48,599,600
02-Oct-01	51.63	53.55	51.56	53.05	40,430,400
01-Oct-01	50.94	52.5	50.41	51.79	34,999,800
28-Sep-01	49.62	51.59	48.98	51.17	58,320,600
27-Sep-01	50.1	50.68	48.0	49.96	40,595,600
26-Sep-01	51.51	51.8	49.55	50.27	29,262,200
25-Sep-01	52.27	53.0	50.16	51.3	42,470,300
24-Sep-01	50.65	52.45	49.87	52.01	42,790,100
21-Sep-01	47.92	50.6	47.5	49.71	92,488,300
20-Sep-01	52.35	52.61	50.67	50.76	58,991,600
19-Sep-01	54.46	54.7	50.6	53.87	63,475,100
18-Sep-01	53.41	55.0	53.17	54.32	41,591,300
17-Sep-01	54.02	55.1	52.8	52.91	63,751,000
10-Sep-01	54.92	57.95	54.7	57.58	42,235,900
07-Sep-01	56.11	57.36	55.31	55.4	44,931,900
06-Sep-01	56.56	58.39	55.9	56.02	56,178,400
05-Sep-01	56.18	58.39	55.39	57.74	44,735,300
04-Sep-01	57.19	59.08	56.07	56.1	33,594,600

(cont.)

Table 8.1 *(cont.)*

Date	Open	High	Low	Close	Volume
31-Aug-01	56.85	58.06	56.3	57.05	28,950,400
30-Aug-01	59.04	59.66	56.52	56.94	48,816,000
29-Aug-01	61.05	61.3	59.54	60.25	24,085,000
28-Aug-01	62.34	62.95	60.58	60.74	23,711,400
27-Aug-01	61.9	63.36	61.57	62.31	22,281,400
24-Aug-01	59.6	62.28	59.23	62.05	31,699,500
23-Aug-01	60.67	61.53	59.0	59.12	25,906,600
22-Aug-01	61.13	61.15	59.08	60.66	39,053,600
21-Aug-01	62.7	63.2	60.71	60.78	23,555,900
20-Aug-01	61.66	62.75	61.1	62.7	24,185,600
17-Aug-01	63.78	64.13	61.5	61.88	26,117,100
16-Aug-01	62.84	64.71	62.7	64.62	21,952,800
15-Aug-01	64.71	65.05	63.2	63.2	19,751,500
14-Aug-01	65.75	66.09	64.45	64.69	18,240,600
13-Aug-01	65.24	65.99	64.75	65.83	16,337,700

9. Valuing by Expected Utility

9.1 Limitations of Arbitrage Pricing

Although arbitrage can be a powerful tool in determining the appropriate cost of an investment, it is more the exception than the rule that it will result in a unique cost. Indeed, as the following example indicates, a unique no-arbitrage option cost will not even result in simple one-period option problems if there are more than two possible next-period security prices.

Example 9.1a Consider the call option example given in Section 5.1. Again, let the initial price of the security be 100, but now suppose that the price at time 1 can be any of the values 50, 200, and 100. That is, we now allow for the possibility that the price of the stock at time 1 is unchanged from its initial price (see Figure 9.1). As in Section 5.1, suppose that we want to price an option to purchase the stock at time 1 for the fixed price of 150.

For simplicity, let the interest rate r equal zero. The arbitrage theorem states that there will be no guaranteed win if there are nonnegative numbers p_{50}, p_{100}, p_{200} that (a) sum to 1 and (b) are such that the expected gains if one purchases either the stock or the option are zero when p_i is the probability that the stock's price at time 1 is i ($i = 50, 100, 200$). Letting G_s denote the gain at time 1 from buying one share of the *stock,* and letting $S(1)$ be the price of that stock at time 1, we have

$$G_s = \begin{cases} 100 & \text{if } S(1) = 200, \\ 0 & \text{if } S(1) = 100, \\ -50 & \text{if } S(1) = 50. \end{cases}$$

Hence,

$$E[G_s] = 100p_{200} - 50p_{50}.$$

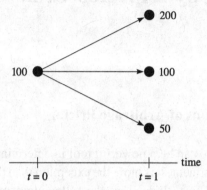

Figure 9.1: Possible Stock Prices at Time 1

Also, if c is the cost of the option, then the gain from purchasing one
option is

$$G_o = \begin{cases} 50 - c & \text{if } S(1) = 200, \\ -c & \text{if } S(1) = 100 \text{ or } S(1) = 50. \end{cases}$$

Therefore,

$$E[G_o] = (50 - c)p_{200} - c(p_{50} + p_{100})$$
$$= 50p_{200} - c.$$

Equating both $E[G_s]$ and $E[G_o]$ to zero shows that the conditions for
the absence of arbitrage are that there exist probabilities and a cost c
such that

$$p_{200} = \tfrac{1}{2}p_{50} \quad \text{and} \quad c = 50p_{200}.$$

Since the leftmost of the preceding equalities implies that $p_{200} \leq 1/3$,
it follows that for any value of c satisfying $0 \leq c \leq 50/3$ we can find
probabilities that make both buying the stock and buying the option fair
bets. Therefore, no arbitrage is possible for any option cost in the interval $[0, 50/3]$. □

9.2   Valuing Investments by Expected Utility

Suppose that you must choose one of two possible investments, each of
which can result in any of n consequences, denoted $C_1, \ldots, C_n$. Suppose

that if the first investment is chosen then consequence i will result with probability p_i $(i = 1, \ldots, n)$, whereas if the second one is chosen then consequence i will result with probability q_i $(i = 1, \ldots, n)$, where $\sum_{i=1}^{n} p_i = \sum_{i=1}^{n} q_i = 1$. The following approach can be used to determine which investment to choose.

We begin by assigning numerical values to the different consequences as follows. First, identify the least and the most desirable consequence, call them c and C respectively; give the consequence c the value 0 and give C the value 1. Now consider any of the other $n - 2$ consequences, say C_i. To value this consequence, imagine that you are given the choice between either receiving C_i or taking part in a random experiment that earns you either consequence C with probability u or consequence c with probability $1 - u$. Clearly your choice will depend on the value of u. If $u = 1$ then the experiment is certain to result in consequence C; since C is the most desirable consequence, you will clearly prefer the experiment to receiving C_i. On the other hand, if $u = 0$ then the experiment will result in the least desirable consequence, namely c, and so in this case you will clearly prefer the consequence C_i to the experiment. Now, as u decreases from 1 down to 0, it seems reasonable that your choice will at some point switch from the experiment to the certain return of C_i, and at that critical switch point you will be indifferent between the two alternatives. Take that indifference probability u as the value of the consequence C_i. In other words, the value of C_i is that probability u such that you are indifferent between either receiving the consequence C_i or taking part in an experiment that returns consequence C with probability u or consequence c with probability $1 - u$. We call this indifference probability the *utility* of the consequence C_i, and we designate it as $u(C_i)$.

In order to determine which investment is superior, we must evaluate each one. Consider the first one, which results in consequence C_i with probability p_i $(i = 1, \ldots, n)$. We can think of the result of this investment as being determined by a two-stage experiment. In the first stage, one of the values $1, \ldots, n$ is chosen according to the probabilities $p_1, \ldots, p_n$; if value i is chosen, you receive consequence C_i. However, since C_i is equivalent to obtaining consequence C with probability $u(C_i)$ or consequence c with probability $1 - u(C_i)$, it follows that the result of the two-stage experiment is equivalent to an experiment in which either consequence C or c is obtained, with C being obtained with probability

$$\sum_{i=1}^{n} p_i u(C_i).$$

Similarly, the result of choosing the second investment is equivalent to taking part in an experiment in which either consequence C or c is obtained, with C being obtained with probability

$$\sum_{i=1}^{n} q_i u(C_i).$$

Since C is preferable to c, it follows that the first investment is preferable to the second if

$$\sum_{i=1}^{n} p_i u(C_i) > \sum_{i=1}^{n} q_i u(C_i).$$

In other words, the value of an investment can be measured by the expected value of the utility of its consequence, and the investment with the largest expected utility is most preferable.

In many investments, the consequences correspond to the investor receiving a certain amount of money. In this case, we let the dollar amount represent the consequence; thus, $u(x)$ is the investor's utility of receiving the amount x. We call $u(x)$ a *utility function*. Thus, if an investor must choose between two investments, of which the first returns an amount X and the second an amount Y, then the investor should choose the first if

$$E[u(X)] > E[u(Y)]$$

and the second if the inequality is reversed, where u is the utility function of that investor. Because the possible monetary returns from an investment often constitute an infinite set, it is convenient to drop the requirement that $u(x)$ be between 0 and 1.

Whereas an investor's utility function is specific to that investor, a general property usually assumed of utility functions is that $u(x)$ is a nondecreasing function of x. In addition, a common (but not universal) feature for most investors is that, if they expect to receive x, then the

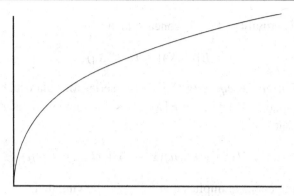

Figure 9.2: A Concave Function

extra utility gained if they are given an additional amount Δ is nonincreasing in x; that is, for fixed $\Delta > 0$, their utility function satisfies

$$u(x + \Delta) - u(x) \text{ is nonincreasing in } x.$$

A utility function that satisfies this condition is called *concave*. It can be shown that the condition of concavity is equivalent to

$$u''(x) \le 0.$$

That is, a function is concave if and only if its second derivative is nonpositive. Figure 9.2 gives the curve of a concave function; such a curve always has the property that the line segment connecting any two of its points always lies below the curve.

An investor with a concave utility function is said to be *risk-averse.* This terminology is used because of the following, known as *Jensen's inequality,* which states that if u is a concave function then, for any random variable X,

$$E[u(X)] \le u(E[X]).$$

Hence, letting X be the return from an investment, it follows from Jensen's inequality that any investor with a concave utility function would prefer the certain return of $E[X]$ to receiving a random return with this mean.

We now give a proof of

Jensen's Inequality   If U is concave then

$$E[U(X)] \leq U(E[X])$$

Proof of Jensen's Inequality.   The Taylor series formula with remainder of $U(x)$ expanded about $\mu = E[X]$ gives, for some value of τ between x and μ, that

$$U(x) = U(\mu) + U'(\mu)(x - \mu) + U''(\tau)(x - \mu)^2/2$$

But U being concave implies that $U'' \leq 0$, showing that

$$U(x) \leq U(\mu) + U'(\mu)(x - \mu)$$

Consequently,

$$U(X) \leq U(\mu) + U'(\mu)(X - \mu)$$

Now take expectations of both sides to obtain the result:

$$E[U(X)] \leq U(\mu) + U'(\mu)E[X - \mu] = U(\mu) \qquad \square$$

An investor with a linear utility function

$$u(x) = a + bx, \quad b > 0,$$

is said to be *risk-neutral* or *risk-indifferent*. For such a utility function,

$$E[u(X)] = a + bE[X]$$

and so it follows that a risk-neutral investor will value an investment only through its expected return.

A commonly assumed utility function is the *log utility function*

$$u(x) = \log(x);$$

see Figure 9.3. Because $\log(x)$ is a concave function, an investor with a log utility function is risk-averse. This is a particularly important utility function because it can be mathematically proven in a variety of situations that an investor faced with an infinite sequence of investments can maximize long-term rate of return by adopting a log utility function and then maximizing the expected utility in each period.

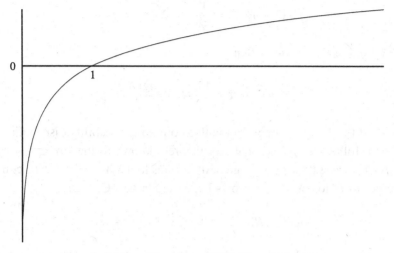

Figure 9.3: A Log Utility Function

To understand why this is true, suppose that the result of each investment is to multiply the investor's wealth by a random amount X. That
is, if W_n denotes the investor's wealth after the nth investment and if X_n
is the nth multiplication factor, then

$$W_n = X_n W_{n-1}, \quad n \geq 1.$$

With W_0 denoting the investor's initial wealth, the preceding implies
that

$$W_n = X_n W_{n-1}$$
$$= X_n X_{n-1} W_{n-2}$$
$$= X_n X_{n-1} X_{n-2} W_{n-3}$$
$$\vdots$$
$$= X_n X_{n-1} \cdots X_1 W_0.$$

If we let R_n denote the rate of return (per investment) from the n investments, then

$$\frac{W_n}{(1 + R_n)^n} = W_0$$

or

$$(1 + R_n)^n = \frac{W_n}{W_0} = X_1 \cdots X_n.$$

Taking logarithms yields that

$$\log(1 + R_n) = \frac{\sum_{i=1}^{n} \log(X_i)}{n}.$$

Now, if the X_i are independent with a common probability distribution, then it follows from a probability theorem known as the *strong law of large numbers* that the average of the values $\log(X_i)$, $i = 1, \ldots, n$, converges to $E[\log(X_i)]$ as n grows larger and larger. Consequently,

$$\log(1 + R_n) \to E[\log(X)] \text{ as } n \to \infty.$$

Therefore, if one has some choice as to the investment – that is, some choice as to the probabilities of the multiplying factors X_i – then the long-run rate of return is maximized by choosing the investment that yields the largest value of $E[\log(X)]$.

Moreover, because $W_n = W_0 X_1 \cdots X_n$, it follows that

$$\log(W_n) = \log(W_0) + \sum_{i=1}^{n} \log(X_i).$$

Hence,

$$E[\log(W_n)] = \log(W_0) + n E[\log(X)]$$

which shows that maximizing $E[\log(X)]$ is equivalent to maximizing the expectation of the log of the final wealth.

The following example shows how much a log utility investor should invest in a favorable gamble.

Example 9.2a An investor with capital x can invest any amount between 0 and x; if y is invested then y is either won or lost, with respective probabilities p and $1 - p$. If $p > 1/2$, how much should be invested by an investor having a log utility function?

Solution. Suppose the amount αx is invested, where $0 \le \alpha \le 1$. Then the investor's final fortune, call it X, will be either $x + \alpha x$ or $x - \alpha x$

with respective probabilities p and $1 - p$. Hence, the expected utility of this final fortune is

$$p \log((1 + \alpha)x) + (1 - p) \log((1 - \alpha)x)$$
$$= p \log(1 + \alpha) + p \log(x) + (1 - p) \log(1 - \alpha) + (1 - p) \log(x)$$
$$= \log(x) + p \log(1 + \alpha) + (1 - p) \log(1 - \alpha).$$

To find the optimal value of α, we differentiate

$$p \log(1 + \alpha) + (1 - p) \log(1 - \alpha)$$

to obtain

$$\frac{d}{d\alpha}(p \log(1 + \alpha) + (1 - p) \log(1 - \alpha)) = \frac{p}{1 + \alpha} - \frac{1 - p}{1 - \alpha}.$$

Setting this equal to zero yields

$$p - \alpha p = 1 - p + \alpha - \alpha p \quad \text{or} \quad \alpha = 2p - 1.$$

Hence, the investor should always invest $100(2p - 1)$ percent of her present fortune. For instance, if the probability of winning is .6 then the investor should invest 20% of her fortune; if it is .7, she should invest 40%. (When $p \leq 1/2$, it is easy to verify that the optimal amount to invest is 0.) □

Our next example adds a time factor to the previous one.

Example 9.2b Suppose in Example 9.2a that, whereas the investment αx must be immediately paid, the payoff of $2\alpha x$ (if it occurs) does not take place until after one period has elapsed. Suppose further that whatever amount is not invested can be put in a bank to earn interest at a rate of r per period. Now, how much should be invested?

Solution. An investor who invests αx and puts the remaining $(1 - \alpha)x$ in the bank will, after one period, have $(1 + r)(1 - \alpha)x$ in the bank, and the investment will be worth either $2\alpha x$ (with probability p) or 0 (with probability $1 - p$). Hence, the expected value of the utility of his fortune is

$$p \log((1+r)(1-\alpha)x + 2\alpha x) + (1-p) \log((1+r)(1-\alpha)x)$$
$$= \log(x) + p \log(1 + r + \alpha - \alpha r)$$
$$+ (1-p) \log(1+r) + (1-p) \log(1-\alpha).$$

Hence, once again the optimal fraction of one's fortune to invest does not depend on the amount of that fortune. Differentiating the previous equation yields

$$\frac{d}{d\alpha}(\text{expected utility}) = \frac{p(1-r)}{1+r+\alpha-\alpha r} - \frac{1-p}{1-\alpha}.$$

Setting this equal to zero and solving yields that the optimal value of α is given by

$$\alpha = \frac{p(1-r) - (1-p)(1+r)}{1-r} = \frac{2p - 1 - r}{1-r}.$$

For instance, if $p = .6$ and $r = .05$ then, although the expected rate of return on the investment is 20% (whereas the bank pays only 5%), the optimal fraction of money to be invested is

$$\alpha = \frac{.15}{.95} \approx .158.$$

That is, the investor should invest approximately 15.8% of his capital and put the remainder in the bank. ☐

Another commonly used utility function is the *exponential utility function*

$$u(x) = 1 - e^{-bx}, \quad b > 0.$$

The exponential is also a risk-averse utility function (see Figure 9.4).

9.3　The Portfolio Selection Problem

Suppose one has the positive amount w to be invested among n different securities. If the amount a is invested in security i ($i = 1, \ldots, n$) then, after one period, that investment returns aX_i, where X_i is a nonnegative random variable. In other words, if we let R_i be the the rate of return from investment i, then

$$a = \frac{aX_i}{1 + R_i} \quad \text{or} \quad R_i = X_i - 1.$$

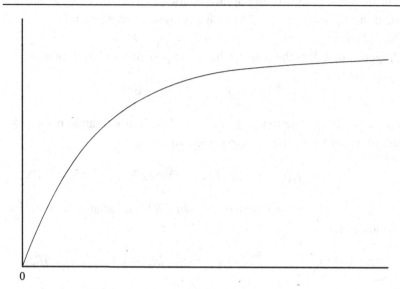

0

Figure 9.4: An Exponential Utility Function

If w_i is invested in each security $i = 1, \ldots, n$, then the end-of-period wealth is

$$W = \sum_{i=1}^{n} w_i X_i.$$

The vector $w_1, \ldots, w_n$ is called a *portfolio*. The problem of determining the portfolio that maximizes the expected utility of one's end-of-period wealth can be expressed mathematically as follows:

choose $w_1, \ldots, w_n$ satisfying

$$w_i \geq 0, \quad i = 1, \ldots, n, \quad \sum_{i=1}^{n} w_i = w,$$

to

maximize $E[U(W)]$,

where U is the investor's utility function for the end-of-period wealth.

To make the preceding problem more tractable, we shall make the assumption that the end-of-period wealth W can be thought of as being a normal random variable. Provided that one invests in many securities that are not too highly correlated, this would appear to be, by the central limit theorem, a reasonable approximation. (It would also be exactly true if the X_i, $i = 1, \ldots, n$, have what is known as a multivariate normal distribution.)

Suppose now that the investor has an exponential utility function

$$U(x) = 1 - e^{-bx}, \quad b > 0,$$

and so the utility function is concave. If Z is a normal random variable, then e^Z is lognormal and has expected value

$$E[e^Z] = \exp\{E[Z] + \text{Var}(Z)/2\}.$$

Hence, as $-bW$ is normal with mean $-bE[W]$ and variance $b^2 \text{Var}(W)$, it follows that

$$E[U(W)] = 1 - E[e^{-bW}] = 1 - \exp\{-bE[W] + b^2 \text{Var}(W)/2\}.$$

Therefore, the investor's expected utility will be maximized by choosing a portfolio that

$$\text{maximizes } E[W] - b\,\text{Var}(W)/2.$$

Observe how this implies that, if two portfolios give rise to random end-of-period wealths W_1 and W_2 such that W_1 has a larger mean and a smaller variance than does W_2, then the first portfolio results in a larger expected utility than does the second. That is,

$$E[W_1] \geq E[W_2] \ \ \& \ \ \text{Var}(W_1) \leq \text{Var}(W_2)$$
$$\implies E[U(W_1)] \geq E[U(W_2)]. \tag{9.1}$$

In fact, provided that all end-of-period fortunes are normal random variables, (9.1) remains valid even when the utility function is not exponential, provided that it is a nondecreasing and concave function. Consequently, if one investment portfolio offers a risk-averse investor an expected return that is at least as large as that offered by a second investment portfolio and with a variance that is no greater than that of the second portfolio, then the investor would prefer the first portfolio.

Let us now compute, for a given portfolio, the mean and variance of W. With security i's rate of return $R_i = X_i - 1$, let

$$r_i = E[R_i], \quad v_i^2 = \text{Var}(R_i).$$

Then, since

$$W = \sum_{i=1}^{n} w_i(1 + R_i) = w + \sum_{i=1}^{n} w_i R_i,$$

we have that

$$E[W] = w + \sum_{i=1}^{n} E[w_i R_i]$$

$$= w + \sum_{i=1}^{n} w_i r_i; \qquad (9.2)$$

$$\mathrm{Var}(W) = \mathrm{Var}\left(\sum_{i=1}^{n} w_i R_i \right)$$

$$= \sum_{i=1}^{n} \mathrm{Var}(w_i R_i)$$

$$+ \sum_{i=1}^{n} \sum_{j \neq i} \mathrm{Cov}(w_i R_i, w_j R_j) \qquad \text{(by Equation (1.11))}$$

$$= \sum_{i=1}^{n} w_i^2 v_i^2 + \sum_{i=1}^{n} \sum_{j \neq i} w_i w_j c(i, j), \qquad (9.3)$$

where

$$c(i, j) = \mathrm{Cov}(R_i, R_j).$$

Example 9.3a An important case which results in W having a normal distribution is the case where $R_1, \ldots, R_n$ has a multivariate normal distribution, defined as follows.

Definition Let $Z_1, \ldots, Z_m$ be independent standard normal random variables. If for some constants $\mu_i, i = 1, \ldots, n$ and $a_{ij}, i = 1, \ldots, n,$ $j = 1, \ldots, m,$

$$X_1 = \mu_1 + a_{11}Z_1 + a_{12}Z_2 + \cdots + a_{1m}Z_m$$
$$X_2 = \mu_2 + a_{21}Z_1 + a_{22}Z_2 + \cdots + a_{2m}Z_m$$
$$\cdots$$
$$X_i = \mu_i + a_{i1}Z_1 + a_{i2}Z_2 + \cdots + a_{im}Z_m$$
$$\cdots$$
$$X_n = \mu_n + a_{n1}Z_1 + a_{n2}Z_2 + \cdots + a_{nm}Z_m$$

we say that $(X_1, \ldots, X_n)$ has a *multivariate normal distribution*.

Because any linear combination $\sum_{i=1}^{n} w_i X_i$ is also a linear combination of the independent normal random variables $Z_1, \ldots, Z_m$, it follows that $\sum_{i=1}^{n} w_i X_i$ is a normal random variable.   □

Example 9.3b Suppose you are thinking about investing your fortune of 100 in two securities whose rates of return have the following expected values and standard deviations:

$$r_1 = .15, \ v_1 = .20; \quad r_2 = .18, \ v_2 = .25.$$

If the correlation between the rates of return is $\rho = -.4$, find the optimal portfolio when employing the utility function

$$U(x) = 1 - e^{-.005x}.$$

Solution. If $w_1 = y$ and $w_2 = 100 - y$, then from Equation (9.2) we obtain

$$E[W] = 100 + .15y + .18(100 - y) = 118 - .03y.$$

Also, since $c(1, 2) = \rho v_1 v_2 = -.02$, Equation (9.3) gives

$$\text{Var}(W) = y^2(.04) + (100 - y)^2(.0625) - 2y(100 - y)(.02)$$
$$= .1425y^2 - 16.5y + 625.$$

We should therefore choose y to maximize

$$118 - .03y - .005(.1425y^2 - 16.5y + 625)/2$$

or, equivalently, to maximize

$$.01125y - .0007125y^2/2.$$

Simple calculus shows that this will be maximized when

$$y = \frac{.01125}{.0007125} = 15.789.$$

That is, the maximal expected utility of the end-of-period wealth is obtained by investing 15.789 in investment 1 and 84.211 in investment 2. Substituting the value $y = 15.789$ into the previous equations gives

$E[W] = 117.526$ and $\text{Var}(W) = 400.006$, with the maximal expected utility being

$$1 - \exp\{-.005(117.526 + .005(400.006)/2)\} = .4416.$$

This can be contrasted with the expected utility of .3904 obtained when all 100 is invested in security 1 or the expected utility of .4413 when all 100 is invested in security 2. □

Example 9.3c Suppose only two securities are under consideration, both with normally distributed returns that have same expected rate of return. Then, since every portfolio will yield the same expected *value*, it follows that the best portfolio for any concave utility function is the one whose end-of-period wealth has minimal variance. If αw is invested in security 1 and $(1 - \alpha)w$ is invested in security 2, then with $c = c(1, 2)$ we have

$$\text{Var}(W) = \alpha^2 w^2 v_1^2 + (1 - \alpha)^2 w^2 v_2^2 + 2\alpha(1 - \alpha)w^2 c$$
$$= w^2[\alpha^2 v_1^2 + (1 - \alpha)^2 v_2^2 + 2c\alpha(1 - \alpha)].$$

Thus, the optimal portfolio is obtained by choosing the value of α that minimizes $\alpha^2 v_1^2 + (1 - \alpha)^2 v_2^2 + 2c\alpha(1 - \alpha)$. Differentiating this quantity and setting the derivative equal to zero yields

$$2\alpha v_1^2 - 2(1 - \alpha)v_2^2 + 2c - 4c\alpha = 0.$$

Solving for α gives the optimal fraction to invest in security 1:

$$\alpha = \frac{v_2^2 - c}{v_1^2 + v_2^2 - 2c}.$$

For instance, suppose the standard deviations of the rate of returns are $v_1 = .20$ and $v_2 = .30$, and that the correlation between the two rates of return is $\rho = .30$. Then, as $c = \rho v_1 v_2 = .018$, we obtain that the optimal fraction of one's investment capital to be used to purchase security 1 is

$$\alpha = \frac{.09 - .018}{.04 + .09 - .036} = 72/94 \approx .766.$$

That is, 76.6% of one's capital should be used to purchase security 1 and 23.4% to purchase security 2.

If the rates of returns are independent, then $c = 0$ and the optimal fraction to invest in security 1 is

$$\alpha = \frac{v_2^2}{v_1^2 + v_2^2} = \frac{1/v_1^2}{1/v_1^2 + 1/v_2^2}.$$

In this case, the optimal percentage of capital to invest in a security is determined by a weighted average, where the weight given to a security is inversely proportional to the variance of its rate of return. This result also remains true when there are n securities whose rates of return are uncorrelated and have equal means. Under these conditions, the optimal fraction of one's capital to invest in security i is

$$\frac{1/v_i^2}{\sum_{j=1}^{n} 1/v_j^2}. \qquad\qquad \Box$$

Determining a portfolio that maximizes the expected utility of one's end-of-period wealth can be computationally quite demanding. Often a reasonable approximation can be obtained when the utility function $U(x)$ satisfies the condition that its second derivative is a nondecreasing function – that is, when

$$U''(x) \text{ is nondecreasing in } x. \qquad\qquad (9.4)$$

It is easily checked that the utility functions

$$U(x) = x^a, \quad 0 < a < 1,$$
$$U(x) = 1 - e^{-bx}, \quad b > 0,$$
$$U(x) = \log(x)$$

all satisfy the condition of Equation (9.4).

We can approximate $U(W)$ by using the first three terms of its Taylor series expansion about the point $\mu = E[W]$. That is, we use the approximation

$$U(W) \approx U(\mu) + U'(\mu)(W - \mu) + U''(\mu)(W - \mu)^2/2.$$

Taking expectations gives that

$$E[U(W)] \approx U(\mu) + U'(\mu)E[W - \mu] + U''(\mu)E[(W - \mu)^2]/2$$
$$= U(\mu) + U''(\mu)v^2/2,$$

where $v^2 = \text{Var}(W)$ and where we have used that

$$E[W - \mu] = E[W] - \mu = \mu - \mu = 0.$$

Therefore, a reasonable approximation to the optimal portfolio is given by the portfolio that maximizes

$$U(E[W]) + U''(E[W]) \, \text{Var}(W)/2. \tag{9.5}$$

If U is a nondecreasing, concave function that also satisfies condition (9.4), then expression (9.5) will have the desired property of being both increasing in $E[W]$ and decreasing in $\text{Var}(W)$.

Utility functions of the form $U(x) = x^a$ or $U(x) = \log(x)$ have the property that there is a vector

$$\alpha_1^*, \dots, \alpha_n^*, \quad \alpha_i^* \geq 0, \quad \sum_{i=1}^{n} \alpha_i^* = 1,$$

such that the optimal portfolio under a specified one of these utility functions is $w\alpha_1^*, \dots, w\alpha_n^*$ for *every* initial wealth w. That is, for these utility functions, the optimal proportion of one's wealth w that should be invested in security i does not depend on w. To verify this, note that

$$W = w \sum_{i=1}^{n} \alpha_i X_i$$

for any portfolio $w\alpha_1, \dots, w\alpha_n$. Hence, if $U(x) = x^a$ then

$$E[U(W)] = E[W^a]$$

$$= E\left[w^a \left(\sum_{i=1}^{n} \alpha_i X_i \right)^a \right]$$

$$= w^a E\left[\left(\sum_{i=1}^{n} \alpha_i X_i \right)^a \right]$$

and so the optimal α_i $(i = 1, \dots, n)$ do not depend on w. (The argument for $U(x) = \log(x)$ is left as an exercise.)

An important feature of the approximation criterion (9.5) is that, when $U(x) = x^a$ $(0 < a < 1)$, the portfolio that maximizes (9.5) also has the property that the percentage of wealth it invests in each security does not depend on w. This follows since equations (9.2) and (9.3) show that, for the portfolio $w_i = \alpha_i w$ $(i = 1, \ldots, n)$,

$$E[W] = wA, \qquad \mathrm{Var}(W) = w^2 B,$$

where

$$A = 1 + \sum_{i=1}^{n} \alpha_i r_i,$$

$$B = \sum_{i=1}^{n} \alpha_i^2 v_i^2 + \sum_{i=1}^{n} \sum_{j \neq i} \alpha_i \alpha_j c(i, j).$$

Thus, since

$$U''(x) = a(a-1)x^{a-2},$$

we see that

$$U(E[W]) + U''(E[W])\,\mathrm{Var}(W)/2$$

$$= w^a A^a + a(a-1)w^{a-2}A^{a-2}w^2 B/2$$

$$= w^a[A^a + a(a-1)A^{a-2}B/2].$$

Therefore, the investment percentages that maximize (9.5) do not depend on w.

Example 9.3d Let us reconsider Example 9.3b, this time using the utility function

$$U(x) = \sqrt{x}.$$

Then, with $\alpha_1 = \alpha$ and $\alpha_2 = 1 - \alpha$ we have

$$A = 1 + .15\alpha + .18(1 - \alpha),$$

$$B = .04\alpha^2 + .0625(1 - \alpha)^2 - 2(.02)\alpha(1 - \alpha),$$

and we must choose the value of α that maximizes

$$f(\alpha) = A^{1/2} - A^{-3/2}B/8.$$

The solution can be obtained by setting the derivative equal to zero and then solving this equation numerically. $\square$

Suppose now that we can invest a positive or negative amount in any investment and, in addition, that all investments are financed by borrowing money at a fixed rate of r per period. If w_i is invested in investment i ($i = 1, \ldots, n$), then the return from this portfolio after one period is

$$R(\mathbf{w}) = \sum_{i=1}^{n} w_i(1 + R_i) - (1 + r) \sum_{i=1}^{n} w_i = \sum_{i=1}^{n} w_i(R_i - r).$$

(If $s = \sum_i w_i$, then s is borrowed from the bank if $s > 0$ and $-s$ is deposited in the bank if $s < 0$.) Let

$$r(\mathbf{w}) = E[R(\mathbf{w})], \qquad V(\mathbf{w}) = \text{Var}(R(\mathbf{w}))$$

and note that

$$r(a\mathbf{w}) = ar(\mathbf{w}), \qquad V(a\mathbf{w}) = a^2 V(\mathbf{w}),$$

where $a\mathbf{w} = (aw_1, \ldots, aw_n)$. Now, let $\mathbf{w}^*$ be such that $r(\mathbf{w}^*) = 1$ and

$$V(\mathbf{w}^*) = \min_{\mathbf{w}:r(\mathbf{w})=1} V(\mathbf{w}).$$

That is, among all portfolios $\mathbf{w}$ whose expected return is 1, the variance of the portfolio's return is minimized under $\mathbf{w}^*$.

We now show that for any $b > 0$, among all portfolios whose expected return is b, the variance of the portfolio's return is minimized under $b\mathbf{w}^*$. To verify this, suppose that $r(\mathbf{y}) = b$. But then

$$r\left(\frac{1}{b}\mathbf{y}\right) = \frac{1}{b}r(\mathbf{y}) = 1,$$

which implies (by the definition of $\mathbf{w}^*$) that

$$V(b\mathbf{w}^*) = b^2 V(\mathbf{w}^*) \le b^2 V\left(\frac{1}{b}\mathbf{y}\right) = V(\mathbf{y}),$$

which completes the verification. Hence, portfolios that minimize the variance of the return are constant multiples of a particular portfolio. This is called the portfolio *separation theorem* because, when analyzing the portfolio decision problem from a mean variance viewpoint, the theorem enables us to separate the portfolio decision problem into a determination of the relative amounts to invest in each investment and the choice of the scalar multiple.

9.3.1   *Estimating Covariances*

In order to create good portfolios, we must first use historical data to estimate the values of $r_i = E[R_i]$, $v_i^2 = \text{Var}(R_i)$, and $c(i, j) = \text{Cov}(R_i, R_j)$ for all i and j. The means r_i and variances v_i^2 can be estimated, as was shown in Section 8.5, by using the sample mean and sample variance of historical rates of return for security i. To estimate the covariance $c(i, j)$ for a fixed pair i and j, suppose we have historical data that covers m periods and let $r_{i,k}$ and $r_{j,k}$ denote (respectively) the rates of return of security i and of security j for period k, $k = 1, \ldots, m$. Then, the usual estimator of

$$\text{Cov}(R_i, R_j) = E[(R_i - r_i)(R_j - r_j)]$$

is

$$\frac{\sum_{k=1}^{m}(r_{i,k} - \bar{r}_i)(r_{j,k} - \bar{r}_j)}{m - 1},$$

where $\bar{r}_i$ and $\bar{r}_j$ are the sample means

$$\bar{r}_i = \frac{\sum_{k=1}^{m} r_{i,k}}{m}, \qquad \bar{r}_j = \frac{\sum_{k=1}^{m} r_{j,k}}{m}.$$

9.4   Value at Risk and Conditional Value at Risk

Let G denote the present value gain from an investment. (If the investment calls for an initial payment of c and returns X after one period, then $G = \frac{X}{1+r} - c$.) The *value at risk* (VAR) of an investment is the value v such that there is only a 1-percent chance that the loss from the investment will be greater than v. Because $-G$ is the loss, the value at risk is the value v such that

$$P\{-G > v\} = .01.$$

The VAR criterion for choosing among different investments, which selects the investment having the smallest VAR, has become popular in recent years.

Example 9.4a Suppose that the gain G from an investment is a normal random variable with mean μ and standard deviation σ. Because

$-G$ is normal with mean $-\mu$ and standard deviation σ, the VAR of this investment is the value of v such that

$$.01 = P\{-G > v\}$$

$$= P\left\{\frac{-G + \mu}{\sigma} > \frac{v + \mu}{\sigma}\right\}$$

$$= P\left\{Z > \frac{v + \mu}{\sigma}\right\},$$

where Z is a standard normal random variable. But from Table 2.1 we see that $P\{Z > 2.33\} = .01$. Therefore,

$$2.33 = \frac{v + \mu}{\sigma}$$

or

$$\text{VAR} = -\mu + 2.33\sigma.$$

Consequently, among investments whose gains are normally distributed, the VAR criterion would select the one having the largest value of $\mu - 2.33\sigma$. □

Remark. The critical value .01 used to define the VAR is the one usually employed because it sets an upper limit to the possible loss that is unlikely to be exceeded. However, an investor might also want to consider other critical values when using the VAR criterion.

The VAR gives a value that has only a 1-percent chance of being exceeded by the loss from an investment. However, rather than choosing the investment having the smallest VAR, it has been suggested that it is better to consider the conditional expected loss, given that it exceeds the VAR. In other words, if the 1-percent event occurs and there is a large loss, then the amount lost will not be the VAR but will be some larger quantity. The conditional expected loss, given that it exceeds the VAR, is called the *conditional value at risk* or CVAR, and the CVAR criterion is to choose the investment having the smallest CVAR.

Example 9.4b If the gain G from an investment is a normal random variable with mean μ and standard deviation σ, then the CVAR is given by

$$\text{CVAR} = E[-G \mid -G > \text{VAR}]$$

$$= E[-G \mid -G > -\mu + 2.33\sigma]$$

$$= E\left[-G \mid \frac{-G + \mu}{\sigma} > 2.33\right]$$

$$= E\left[\sigma\left(\frac{-G + \mu}{\sigma}\right) - \mu \mid \frac{-G + \mu}{\sigma} > 2.33\right]$$

$$= \sigma E\left[\frac{-G + \mu}{\sigma} \mid \frac{-G + \mu}{\sigma} > 2.33\right] - \mu$$

$$= \sigma E[Z \mid Z > 2.33] - \mu,$$

where Z is a standard normal. It can be shown that, for a standard normal random variable Z,

$$E[Z \mid Z > a] = \frac{1}{\sqrt{2\pi}\, P\{Z \geq a\}} e^{-a^2/2}. \tag{9.6}$$

Hence we obtain that

$$\text{CVAR} = \sigma \frac{100}{\sqrt{2\pi}} \exp\{-(2.33)^2/2\} - \mu = 2.64\sigma - \mu.$$

Therefore, the CVAR, which attempts to maximize $\mu - 2.64\sigma$, gives a little more weight to the variance than does the VAR.   □

To verify Equation (9.6), use that the conditional density of Z given that $Z > a$ is

$$f_{Z \mid Z > a}(x) = \frac{1}{\sqrt{2\pi}} \frac{e^{-x^2/2}}{P(Z > a)}, \quad x > a.$$

This gives

$$E[Z \mid Z > a] = \frac{1}{\sqrt{2\pi}\, P(Z > a)} \int_a^\infty x e^{-x^2/2} dx$$

$$= \frac{1}{\sqrt{2\pi}\, P(Z > a)} e^{-a^2/2}$$

9.5   The Capital Assets Pricing Model

The Capital Assets Pricing Model (CAPM) attempts to relate R_i, the one-period rate of return of a specified security i, to R_m, the one-period rate of return of the entire market (as measured, say, by the Standard and Poor's index of 500 stocks). If r_f is the risk-free interest rate (usually taken to be the current rate of a U.S. Treasury bill) then the model assumes that, for some constant β_i,

$$R_i = r_f + \beta_i(R_m - r_f) + e_i,$$

where e_i is a normal random variable with mean 0 that is assumed to be independent of R_m. Letting the expected values of R_i and R_m be r_i and r_m (resp.), the CAPM model (which treats r_f as a constant) implies that

$$r_i = r_f + \beta_i(r_m - r_f)$$

or, equivalently, that

$$r_i - r_f = \beta_i(r_m - r_f).$$

That is, the difference between the expected rate of return of the security and the risk-free interest rate is assumed to equal β_i times the difference between the expected rate of return of the market and the risk-free interest rate. Thus, for instance, if $\beta_i = 1$ (resp. $\frac{1}{2}$ or 2) then the expected amount by which the rate of return of security i exceeds r_f is the same as (resp. one-half or twice) the expected amount by which the overall market's rate of return exceeds r_f. The quantity β_i is known as the *beta* of security i.

Using the linearity property of covariances – along with the result that the covariance of a random variable and a constant is 0 – we obtain from the CAPM that

$$\operatorname{Cov}(R_i, R_m) = \beta_i \operatorname{Cov}(R_m, R_m) + \operatorname{Cov}(e_i, R_m)$$

$$= \beta_i \operatorname{Var}(R_m) \quad \text{(since } e_i \text{ and } R_m \text{ are independent).}$$

Therefore, letting $v_m^2 = \operatorname{Var}(R_m)$, we see that

$$\beta_i = \frac{\operatorname{Cov}(R_i, R_m)}{v_m^2}.$$

Example 9.5a   Suppose that the current risk-free interest rate is 6% and that the expected value and standard deviation of the market rate of return are .10 and .20, respectively. If the covariance of the rate of return of a given stock and the market's rate of return is .05, what is the expected rate of return of that stock?

Solution. Since

$$\beta = \frac{.05}{(.20)^2} = 1.25,$$

it follows (assuming the validity of the CAPM) that

$$r_i = .06 + 1.25(.10 - .06) = .11.$$

That is, the stock's expected rate of return is 11%.   □

If we let $v_i^2 = \text{Var}(R_i)$ then under the CAPM it follows, using the assumed independence of R_m and e_i, that

$$v_i^2 = \beta_i^2 v_m^2 + \text{Var}(e_i).$$

If we think of the variance of a security's rate of return as constituting the risk of that security, then the foregoing equation states that the risk of a security is the sum of two terms: the first term, $\beta_i^2 v_m^2$, is called the *systematic risk* and is due to the combination of the security's beta and the inherent risk in the market; the second term, $\text{Var}(e_i)$, is called the *specific risk* and is due to the specific stock being considered.

9.6   Rates of Return: Single-Period and Geometric Brownian Motion

Let $S_i(t)$ be the price of security i at time t ($t \geq 0$), and assume that these prices follow a geometric Brownian motion with drift parameter μ_i and volatility parameter σ_i. If R_i is the one-period rate of return for security i, then

$$\frac{S_i(1)}{1 + R_i} = S_i(0)$$

or, equivalently,

$$R_i = \frac{S_i(1)}{S_i(0)} - 1.$$

Since $S_i(1)/S_i(0)$ has the same probability distribution as e^X when X is a normal random variable with mean μ_i and variance σ_i^2, it follows that

$$r_i = E[R_i] = E\left[\frac{S_i(1)}{S_i(0)}\right] - 1$$

$$= E[e^X] - 1$$

$$= \exp\{\mu_i + \sigma_i^2/2\} - 1.$$

Also,

$$v_i^2 = \text{Var}(R_i) = \text{Var}\left(\frac{S_i(1)}{S_i(0)}\right)$$

$$= \text{Var}(e^X)$$

$$= E[e^{2X}] - (E[e^X])^2$$

$$= \exp\{2\mu_i + 2\sigma_i^2\} - (\exp\{\mu_i + \sigma_i^2/2\})^2$$

$$= \exp\{2\mu_i + 2\sigma_i^2\} - \exp\{2\mu_i + \sigma_i^2\},$$

where the next-to-last equality used the fact that $2X$ is normal with mean $2\mu_i$ and variance $4\sigma_i^2$ to determine $E[e^{2X}]$.

Thus, the expected one-period rate of return is $\exp\{\mu_i + \sigma_i^2/2\} - 1$; note that this is *not* the expected value of the average spot rate of return by time 1. For if we let $\bar{R}_i(t)$ be the average spot rate of return by time t (i.e., the yield curve), then

$$\frac{S_i(t)}{S_i(0)} = e^{t\bar{R}_i(t)},$$

implying that

$$\bar{R}_i(t) = \frac{1}{t}\log\left(\frac{S_i(t)}{S_i(0)}\right).$$

Since $\log(S_i(t)/S_i(0))$ is a normal random variable with mean $\mu_i t$ and variance $t\sigma_i^2$, it follows that $\bar{R}_i(t)$ is a normal random variable with

$$E[\bar{R}_i(t)] = \mu_i, \qquad \text{Var}(\bar{R}_i(t)) = \sigma_i^2/t.$$

Thus, the expected value and variance of the one-period yield function for geometric Brownian motion are its parameters μ_i and σ_i^2.

9.7   Exercises

Exercise 9.1   The utility function of an investor is $u(x) = 1 - e^{-x}$. The investor must choose one of two investments. If his fortune after investment 1 is a random variable with density function $f_1(x) = e^{-x}$, $x > 0$,   and his fortune after investment 2 is a random variable with density function $f_2(x) = 1/2$, $0 < x < 2$,   which investment should he choose?

Exercise 9.2   If an individual invests the amount a, then the return from that investment is aX, where

$$P(X = -1) = 0.4, \quad P(X = 0.2) = 0.5, \quad P(X = 2.5) = 0.1$$

What is the optimal value of a for a risk-averse individual?

Exercise 9.3   In Example 9.2a, show that if $p \leq 1/2$ then the optimal amount to invest is 0.

Exercise 9.4   In Example 9.2b, show that if $p \leq 1/2$ then the optimal amount to invest is 0.

Exercise 9.5   Suppose in Example 9.3b that $\rho = 0$. What is the optimal portfolio?

Exercise 9.6   Suppose in Example 9.3b that $r_1 = .16$. Determine the maximal expected utility and compare it with (a) the expected utility obtained when everything is invested in security 1 and (b) the expected utility obtained when everything is invested in security 2.

Exercise 9.7   Show that the percentage of one's wealth that should be invested in each security when attempting to maximize $E[\log(W)]$ does not depend on the amount of initial wealth.

Exercise 9.8   Verify that $U''(x)$ is nondecreasing in x when $x > 0$ and when

(a) $U(x) = x^a, 0 < a < 1$;
(b) $U(x) = 1 - e^{-bx}, b > 0$;
(c) $U(x) = \log(x)$.

Exercise 9.9 Does the percentage of one's wealth to be invested in each security when attempting to maximize the approximation (9.5) depend on initial wealth when $U(x) = \log(x)$?

Exercise 9.10 Use the approximation to $E[U(W)]$ given by (9.5) to determine the optimal amounts to invest in each security in Example 9.3a when using the utility function $U(x) = 1 - e^{-.005x}$. Compare your results with those obtained in that example.

Exercise 9.11 Suppose we want to choose a portfolio with the objective of maximizing the probability that our end-of-period wealth be at least g, where $g > w$. Assuming that W is normal, the optimal portfolio will be the one that maximizes what function of $E[W]$ and $\mathrm{Var}(W)$?

Exercise 9.12 Find the optimal portfolio in Example 9.3b if your objective is to maximize the probability that your end-of-period wealth be at least: (a) 110; (b) 115; (c) 120; (d) 125. Assume normality.

Exercise 9.13 Find the solution of Example 9.3d.

Exercise 9.14 If the beta of a stock is .80, what is the expected rate of return of that stock if the expected value of the market's rate of return is .07 and the risk-free interest rate is 5%? What if the risk-free interest rate is 10%? Assume the CAPM.

Exercise 9.15 If β_i is the beta of stock i for $i = 1, \ldots, k$, what would be the beta of a portfolio in which α_i is the fraction of one's capital that is used to purchase stock i $(i = 1, \ldots, k)$?

Exercise 9.16 A single-factor model supposes that R_i, the one-period rate of return of a specified security, can be expressed as

$$R_i = a_i + b_i F + e_i,$$

where F is a random variable (called the "factor"), e_i is a normal random variable with mean 0 that is independent of F, and a_i and b_i are constants that depend on the security. Show that the CAPM is a single-factor model, and identify a_i, b_i, and F.

Exercise 9.17   Let X_1 and X_2 be independent normal random variables, both with mean 1 and variance 1. Investor A has a strictly concave utility function.

(a) Is it possible to tell whether A would prefer a final fortune of 2 or a final fortune of $X_1 + X_2$?

(b) Is it possible to tell whether A would prefer a final fortune of $2X_1$ or a final fortune of $X_1 + X_2$?

(c) Is it possible to tell whether A would prefer a final fortune of $3X_1$ or a final fortune of $X_1 + X_2$?

(d) If A's utility function is $u(x) = 1 - e^{-x}$, which final fortune in part (c) is preferable?

Exercise 9.18   If $X_1, \ldots, X_n$ has the multivariate normal distribution with parameters as given in Example 9.3a, show that

$$\text{Cov}(X_i, X_j) = \sum_{r=1}^{n} a_{ir} a_{jr}.$$

REFERENCES

References [2], [3], and [5] deal with utility theory.

[1] Breiman, L. (1960). "Investment Policies for Expanding Businesses Optimal in a Long Run Sense." *Naval Research Logistics Quarterly* 7: 647–51.

[2] Ingersoll, J. E. (1987). *Theory of Financial Decision Making.* Lanham, MD: Rowman & Littlefield.

[3] Pratt, J. (1964). "Risk Aversion in the Small and in the Large." *Econometrica* 32: 122–30.

[4] Thorp, E. O. (1975). "Portfolio Choice and the Kelly Criterion." In W. T. Ziemba and R. G. Vickson (Eds.), *Stochastic Optimization Models in Finance.* New York: Academic Press.

[5] von Neumann, J., and O. Morgenstern (1944). *Theory of Games and Economic Behavior.* Princeton, NJ: Princeton University Press.

10. Stochastic Order Relations

10.1   First-Order Stochastic Dominance

Of random variables X and Y, we say that X *stochastically dominates* Y, (or equivalently, that X is *stochastically larger* than Y), written as $X \geq_{st} Y$, if for all t

$$P(X > t) \geq P(Y > t)$$

That is, $X \geq_{st} Y$ if for evey constant t, it is at least as likely that X will exceed t as it is that Y will.

Remark. Because a probability is always a continuous function on events, an equivalent definition would be that $X \geq_{st} Y$ if $P(X \geq t) \geq P(Y \geq t)$ for all t.

The following proposition gives an equivalent condition.

Proposition 10.1.1 $X \geq_{st} Y$ *if and only if* $E[h(X)] \geq E[h(Y)]$ *for all increasing functions* h.

Our proof uses two lemmas.

Lemma 10.1.1   *If X is a nonnegative random variable, then*

$$E[X] = \int_0^\infty P(X > t)\, dt$$

Proof.   For $t > 0$, define the random variable $I(t)$ by

$$I(t) = \begin{cases} 1, & \text{if } t < X \\ 0, & \text{if } t \geq X \end{cases}$$

Now,

$$\int_0^\infty I(t)\, dt = \int_0^X I(t)\, dt + \int_X^\infty I(t)\, dt = X$$

Consequently,

$$E[X] = E\left[\int_0^\infty I(t)\, dt\right] = \int_0^\infty E[I(t)]\, dt = \int_0^\infty P(X > t)\, dt \quad \square$$

Lemma 10.1.2 *If $X \geq_{st} Y$; then $E[X] \geq E[Y]$.*

Proof. Suppose first that X and Y are nonnegative random variables. Then Lemma 10.1.1 and the stochastic dominance definition give

$$E[X] = \int_0^\infty P(X > t)\, dt \geq \int_0^\infty P(Y > t)\, dt = E[Y]$$

Hence, the result is true when the random variables are nonnegative.

To prove the result in general, note that any number a can be expressed as the difference of its positive and negative parts:

$$a = a^+ - a^-$$

where

$$a^+ = \max(a, 0), \quad a^- = \max(-a, 0)$$

The preceding follows because if $a \geq 0$, then $a^+ = a$ and $a^- = 0$; whereas if $a < 0$, then $a^+ = 0$ and $a^- = -a$. So, assume that $X \geq_{st} Y$ and express X and Y as the difference of their positive and negative parts:

$$X = X^+ - X^-, \quad Y = Y^+ - Y^-$$

Now, for any $t \geq 0$,

$$P(X^+ > t) = P(X > t)$$

$$\geq P(Y > t) \quad \text{(because } X \geq_{st} Y\text{)}$$

$$= P(Y^+ > t)$$

and

$$P(X^- > t) = P(-X > t)$$

$$= P(X < -t)$$

$$\leq P(Y < -t) \quad \text{(because } X \geq_{st} Y)$$

$$= P(-Y > t)$$

$$= P(Y^- > t)$$

Hence, $X^+ \geq_{st} Y^+$ and $X^- \leq_{st} Y^-$. As these random variables are all nonnegative, we have that $E[X^+] \geq E[Y^+]$ and that $E[X^-] \leq E[Y^-]$. The result now follows because

$$E[X] = E[X^+] - E[X^-] \geq E[Y^+] - E[Y^-] = E[Y] \qquad \square$$

We are now ready to prove Proposition 10.1.1.

Proof of Proposition 10.1.1 Suppose that $X \geq_{st} Y$ and that h is a increasing function. To show that $E[h(X)] \geq E[h(Y)]$, we first show that $h(X) \geq_{st} h(Y)$. Now, for any t, because h is increasing it follows that there is some value – call it $h^{-1}(t)$ – such that the event that $h(X) > t$ is equivalent either to the event that $X \geq h^{-1}(t)$ or to the event that $X > h^{-1}(t)$. (If there is a unique value y such that $h(y) = t$, then the latter case holds and $y = h^{-1}(t)$.) Assuming the latter case, we have

$$P(h(X) > t) = P(X > h^{-1}(t))$$

$$\geq P(Y > h^{-1}(t))$$

$$= P(h(Y) > t)$$

Because a similar argument would hold if $h(X) > t$ were equivalent to $X \geq h^{-1}(t)$, it follows that $h(X) \geq_{st} h(Y)$. Lemma 10.1.2 now gives that $E[h(X)] \geq E[h(Y)]$.

To go the other way, assume that $E[h(X)] \geq E[h(Y)]$ for all increasing functions h. Now, for fixed t, define the function h_t by

$$h_t(x) = \begin{cases} 0, & \text{if } x \leq t \\ 1, & \text{if } x > t \end{cases}$$

Then $h_t(x)$ is increasing, and so

$$E[h_t(X)] \geq E[h_t(Y)]$$

But $E[h_t(X)] = P(X > t)$ and $E[h_t(Y)] = P(Y > t)$, thus showing that $X \geq_{st} Y$. $\qquad\qquad\qquad\qquad\qquad\qquad\qquad\qquad\quad$ □

10.2   Using Coupling to Show Stochastic Dominance

One way to show that $X \geq_{st} Y$ is to find random variables X' and Y' such that X' has the same distribution as X and Y' has the same distribution as Y, which are such that it is always the case that $X' \geq Y'$. For assume that we have found such random variables. Then, because $Y' > t$ implies that $X' > t$, it follows that

$$P(Y' > t) \leq P(X' > t)$$

which, because $P(X' > t) = P(X > t)$ and $P(Y' > t) = P(Y > t)$, proves that $X \geq_{st} Y$. This approach to establishing that one random variable is stochastically larger than another is called *coupling*.

Example 10.2a   Show that a Poisson random variable is stochastically increasing in its mean. That is, show that a Poisson random variable with mean $\lambda_1 + \lambda_2$ is stochastically larger than a Poisson random variable with mean λ_1 when $\lambda_i > 0, i = 1, 2$.

Solution.   For a Poisson random variable X with mean λ,

$$P(X \geq j) = \sum_{i=j}^{\infty} e^{-\lambda}\lambda^i / i!$$

However, it is not easy to directly verify that the preceding is an increasing function of λ for any j. An easier solution is obtained by coupling. Let X_1 and X_2 be independent Poisson random variables, with X_i having mean $\lambda_i, i = 1, 2$. Then, using that the sum of independent Poisson random variables is also Poisson, it follows that $X_1 + X_2$ is Poisson with mean $\lambda_1 + \lambda_2$. Because $X_1 + X_2 \geq X_1$, the result follows. $\qquad$ □

It turns out that if $X \geq_{st} Y$, then it is always possible to find random variables X' and Y' such that X' has the same distribution as X, Y' has the same distribution as Y, and $X' \geq Y'$. We give a proof of this result when X and Y are continuous random variables. We start with a lemma of independent interest.

Lemma 10.2.1 *If F is a continuous distribution function and U a uniform $(0, 1)$ random variable, then the random variable $F^{-1}(U)$ has distribution function F, where $F^{-1}(u)$ is defined to be that value such that $F(F^{-1}(u)) = u$.*

Proof. Because a distribution function is increasing, it follows that the inequalities $a \leq x$ and $F(a) \leq F(x)$ are equivalent. Hence,

$$P(F^{-1}(U) \leq x) = P(F(F^{-1}(U)) \leq F(x))$$

$$= P(U \leq F(x))$$

$$= F(x) \qquad \square$$

Proposition 10.2.1 *If $X \geq_{st} Y$, then there are random variables X' having the same distribution as X, and Y' having the same distribution as Y, such that $X' \geq Y'$.*

Solution. Assume that X and Y are continuous, with respective distribution functions F and G, and that $X \geq_{st} Y$. Because $X \geq_{st} Y$ means that $F(x) \leq G(x)$ for all x, it follows that

$$F(G^{-1}(u)) \leq G(G^{-1}(u)) = u = F(F^{-1}(u))$$

Because F is increasing, the preceding shows that $G^{-1}(u) \leq F^{-1}(u)$. Now, let U be a uniform $(0, 1)$ random variable and set $X' = F^{-1}(U)$ and $Y' = G^{-1}(U)$. The preceding gives that $X' \geq Y'$, and the result follows from Lemma 10.2.1. $\qquad \square$

The following is a useful result, which is easily established by a coupling argument.

Theorem 10.2.1 *Let $X_1, \ldots, X_n$ and $Y_1, \ldots, Y_n$ be vectors of independent random variables, and suppose that $X_i \geq_{st} Y_i$ for each $i = 1, \ldots, n$. Show that $g(X_1, \ldots, X_n) \geq_{st} g(Y_1, \ldots, Y_n)$ whenever $g(x_1, \ldots, x_n)$ is increasing in each component.*

Proof. Let $g(x_1, \ldots, x_n)$ be an increasing function. Let F_i be the distribution function of X_i and let G_i be the distribution function of Y_i, for $i = 1, \ldots, n$. Let $U_1, \ldots, U_n$ be independent uniform $(0, 1)$ random variables, and set

$$X_i' = F_i^{-1}(U_i), \quad Y_i' = G_i^{-1}(U_i), \quad i = 1, \ldots, n$$

Because $X_i' \geq Y_i'$ for all i, it follows that $g(X_1', \ldots, X_n') \geq g(Y_1', \ldots, Y_n')$. The result now follows because $g(X_1', \ldots, X_n')$ has the same distribution as $g(X_1, \ldots, X_n)$ and $g(Y_1', \ldots, Y_n')$ has the same distribution as $g(Y_1, \ldots, Y_n)$. $\square$

10.3   Likelihood Ratio Ordering

Assume that the random variables X and Y are continuous random variables, with X having density function f and Y having density function g. We say that X *is likelihood ratio larger* than Y if $\frac{f(x)}{g(x)}$ is increasing in x over the region where either $f(x)$ or $g(x)$ is greater than 0.

Similarly, if X and Y are discrete random variables, we say that X *is likelihood ratio larger* than Y if $\frac{P(X=x)}{P(Y=x)}$ is increasing in x over the region where either $P(X = x)$ or $P(Y = x)$ is greater than 0.

We now show that likelihood ratio ordering is stronger than stochastic order.

Proposition 10.3.1 *If X is likelihood ratio larger than Y, then X is stochastically larger than Y.*

Proof. Suppose X and Y have respective probability density (or mass) functions f and g, and suppose that $\frac{f(x)}{g(x)} \uparrow x$. For any a, we need to show that

$$\int_{x>a} f(x)dx \geq \int_{x>a} g(x)dx$$

(The preceding integrals should be interpreted as sums when X and Y are discrete.) There are two cases:

Case 1: $f(a) \geq g(a)$

Here, if $x > a$ then $\frac{f(x)}{g(x)} \geq \frac{f(a)}{g(a)} \geq 1$. Hence, $f(x) \geq g(x)$ when $x \geq a$, giving the result.

Case 2: $f(a) < g(a)$

Here, if $x \leq a$ then $\frac{f(x)}{g(x)} \leq \frac{f(a)}{g(a)} < 1$, giving that $\int_{x \leq a} f(x)\, dx < \int_{x \leq a} g(x)\, dx$, which implies the result on subtracting both sides of this inequality from 1. $\square$

Example 10.3a   Let X be a random variable with density function $f(x)$. The density function f_t given by

$$f_t(x) = Ce^{tx} f(x)$$

where $C^{-1} = \int e^{ty} f(y)dy$, is said to be a tilted density with regard to f. Because

$$\frac{f_t(x)}{f(x)} = \frac{e^{tx}}{\int e^{ty} f(y)dy}$$

is increasing in x when $t > 0$ and decreasing when $t < 0$, it follows that a random variable X_t having density function f_t is likelihood ratio (and thus also stochastically) larger than X when $t > 0$ and likelihood ratio (and thus also stochastically) smaller when $t < 0$. $\square$

10.4   A Single-Period Investment Problem

Consider a situation in which one has an initial fortune w and must decide on an amount $y, 0 \leq y \leq w$, to invest. Suppose that an investment of size y returns the amount $yX + (1 + r)(w - y)$ at the end of one period, where X is a nonnegative random variable having a known distribution and r is a specified interest rate earned by the uninvested amount. Furthermore, suppose that, for a given increasing, concave utility function u, the objective is to maximize the expected utility of the end-of-period wealth. That is, with $\beta = 1 + r$, the objective is to find

$$M = \max_{0 \leq y \leq w} E[u(yX + \beta(w - y))]$$

Now, suppose that X is a continuous random variable having density function f. Then

$$M = \max_{0 \leq y \leq w} E[u((X - \beta)y + \beta w)]$$

$$= \max_{0 \leq y \leq w} \int_{-\infty}^{\infty} u((x - \beta)y + \beta w) f(x)\, dx$$

Differentiating the term inside the maximum yields that

$$\frac{d}{dy} \int_0^\infty u((x - \beta)y + \beta w) f(x) \, dx$$

$$= \int_0^\infty u'((x - \beta)y + \beta w)(x - \beta) f(x) \, dx$$

$$= \int_0^\infty h(y, x) f(x) \, dx$$

where

$$h(y, x) = u'((x - \beta)y + \beta w)(x - \beta)$$

Setting the preceding derivative equal to 0 shows that the maximizing value of y, call if y_f, is such that

$$\int_0^\infty h(y_f, x) f(x) \, dx = 0 \qquad (10.1)$$

The following properties of $h(y, x)$ will be needed in the sequel.

Lemma 10.4.1 *For fixed x, $h(y, x)$ is decreasing in y. In addition,*

$$h(y, x) \le 0 \quad \text{if} \quad x \le \beta$$
$$h(y, x) \ge 0 \quad \text{if} \quad x \ge \beta$$

Proof.

Case 1: $x \le \beta$

That $h(x, y) \downarrow y$ follows in this case by the the following string of implications:

$$x \le \beta \Rightarrow (x - \beta)y + \beta w \downarrow y$$
$$\Rightarrow u'((x - \beta)y + \beta w) \uparrow y \quad (\text{because } u \text{ concave} \Rightarrow u'(v) \downarrow v)$$
$$\Rightarrow h(y, x) = (x - \beta) u'((x - \beta)y + \beta w) \downarrow y$$

Also, $h(y, x) \le 0$ because $x - \beta \le 0$ and $u' \ge 0$ (because u is increasing).

Case 2: $x \geq \beta$

$$x \geq \beta \Rightarrow (x - \beta)y + \beta w \uparrow y$$
$$\Rightarrow u'((x - \beta)y + \beta w) \downarrow y \quad \text{(because } u'(v) \downarrow v\text{)}$$
$$\Rightarrow h(y, x) = (x - \beta) u'((x - \beta)y + \beta w) \downarrow y$$

Moreover, $h(y, x)$, being the product of two nonnegative factors, is nonnegative. $\square$

Now consider two scenarios for an investor with initial wealth w: one where the multiplicative random variable is X_1 and and the second where the multiplicative random variable is X_2, where X_1 has density function f and X_2 has density function g. Under what conditions on f and g would the optimal amount invested in the first scenario always be at least as large as the optimal amount invested in the second scenario for every increasing, concave utility function? That is, when is $y_f \geq y_g$? Although one might initially guess that it would be sufficient for X_1 to be stochastically larger than X_2, that this is not the case is shown by the following example.

Example 10.4a Suppose the utility function is

$$u(x) = \begin{cases} x, & \text{if } x \leq 100 \\ 100, & \text{if } x > 100 \end{cases}$$

If we suppose that

$$P(X_1 = 4) = P(X_1 = 0) = 1/2$$

whereas

$$P(X_2 = 3) = P(X_2 = 0) = 1/2$$

then it is easy to check that X_1 is stochastically larger than X_2. Further, suppose that the initial wealth is $w = 30$ and that the interest rate is $r = 0$. Because the utility function is flat at values of 100 or larger, the optimal amount to invest in the X_1 factor problem cannot exceed 70/3 because investing more than 70/3 would yield the same utility value (of 100) as investing 70/3 if $X_1 = 4$ and a smaller utility if $X_1 = 0$. On the other hand, it is easy to check that the optimal amount to invest in the X_2 factor problem is 30. $\square$

Thus we see from Example 10.1 that having a stochastically larger investment return factor does not necessarily imply that a larger amount should be invested. This result is, however, true when the investment returns are likelihood ratio ordered.

Theorem 10.4.1 *If f and g are density functions of nonnegative random variables, for which $\frac{f(x)}{g(x)}$ increases in x, then $y_f \geq y_g$. That is, when f is a likelihood ratio ordered larger density than g, then the optimal amount to invest when the multiplicative factor has density f is larger than when it has density g.*

Proof. From Equation (10.1), the optimal amount to invest when X has density g, namely, y_g, satisfies

$$\int_0^\infty h(y_g, x) g(x)\, dx = 0$$

We want to show that if $\frac{f(x)}{g(x)} \uparrow x$, then $y_f \geq y_g$, where y_f is such that

$$\int_0^\infty h(y_f, x) f(x)\, dx = 0$$

Because $h(y, x)$ is decreasing in y (Lemma 10.4.1), it follows that the inequality $y_f \geq y_g$ is equivalent to the inequality $\int_0^\infty h(y_g, x) f(x)\, dx \geq \int_0^\infty h(y_f, x) f(x)\, dx$. Thus, it suffices to prove that

$$\int_0^\infty h(y_g, x) f(x)\, dx \geq 0$$

Now,

$$\int_0^\infty h(y_g, x) f(x)\, dx = \int_0^\beta h(y_g, x) f(x)\, dx + \int_\beta^\infty h(y_g, x) f(x)\, dx$$

If $x \leq \beta$, then $\frac{f(x)}{g(x)} \leq \frac{f(\beta)}{g(\beta)}$, giving that $f(x) \leq \frac{f(\beta)}{g(\beta)} g(x)$. Also, if $x \leq \beta$, then, from Lemma 10.4.1, $h(y_g, x) \leq 0$. Hence,

$$\int_0^\beta h(y_g, x) f(x)\, dx \geq \frac{f(\beta)}{g(\beta)} \int_0^\beta h(y_g, x) g(x)\, dx \qquad (10.2)$$

If $x \geq \beta$, then $\frac{f(x)}{g(x)} \geq \frac{f(\beta)}{g(\beta)}$ and, by Lemma 10.4.1, $h(y_g, x) \geq 0$. Hence,

$$\int_\beta^\infty h(y_g, x) f(x)\, dx \geq \frac{f(\beta)}{g(\beta)} \int_\beta^\infty h(y_g, x) g(x)\, dx \qquad (10.3)$$

Thus, by (10.2) and (10.3), we obtain

$$\int_0^\infty h(y_g, x) f(x)\, dx \geq \frac{f(\beta)}{g(\beta)} \int_0^\infty h(y_g, x) g(x)\, dx = 0$$

and the result is proven. $\square$

10.5   Second-Order Dominance

Whereas X stochastically dominates Y requires that $E[h(X)] \geq E[h(Y)]$ for all increasing functions h, we often are interested in conditions under which the preceding is required to hold not for all increasing functions h but only for those increasing functions that are also concave. That is, we are interested in when a final fortune of X is always preferable to a final fortune of Y provided that the investor has an increasing concave utility function.

Definition. We say that X *second order dominates* Y, written as $X \geq_{icv} Y$, if

$$E[h(X)] \geq E[h(Y)] \quad \text{for all functions } h \text{ that}$$
$$\text{are both increasing and concave}$$

Remarks.

1. The notation $X \geq_{icv} Y$ is used because equivalent terminology to X second-order dominating Y is that X is stochastically larger than Y in the increasing, concave sense.
2. If X has expected value $E[X]$, then it follows from Jensen's inequality (see Section 9.2) that the constant random variable $E[X]$ second order dominates X.

For a specified value of a, let the function h_a be defined as follows:

$$h_a(x) = \begin{cases} x, & \text{if } x \leq a \\ a, & \text{if } x > a \end{cases}$$

Because $h_a(x)$ is an increasing straight line that becomes flat when it hits a, it is an increasing, concave function. Writing

$$h_a(X) = a - (a - h_a(X))$$

we obtain, on applying Lemma 10.1.1 to the nonnegative random variable $a - h_a(X)$, that

$$E[h_a(X)] = a - E[a - h_a(X)]$$

$$= a - \int_0^\infty P(a - h_a(X) > t)\, dt$$

$$= a - \int_0^\infty P(h_a(X) < a - t)\, dt$$

$$= a - \int_0^\infty P(X < a - t)\, dt$$

$$= a - \int_{-\infty}^a P(X < y)\, dy$$

It follows from the preceding that if X second-order stochastically dominates Y then

$$\int_{-\infty}^a P(X < y)\, dy \le \int_{-\infty}^a P(Y < y)\, dy \quad \text{for all } a \qquad (10.4)$$

In fact, it can be shown that the preceding is also a sufficient condition for $X \ge_{icv} Y$. That is, the following theorem holds.

Theorem 10.5.1 *X second-order stochastically dominates Y if and only if (10.4) holds.*

   Although the preceding theorem gives a necessary and sufficient condition for one random variable to second-order dominate another, we will not make use of it in considering second-order dominance among normal random variables.

10.5.1   *Normal Random Variables*

This subsection is concerned with showing that a normal random variable is increasing in its mean and decreasing in its variance in the second order stochastic dominance sense. That is, the following holds.

Theorem 10.5.2   *If* X_i, $i = 1, 2$, *are normal random variables with respective means* μ_i *and variances* σ_i^2, *then*

$$\mu_1 \geq \mu_2, \sigma_1 \leq \sigma_2 \Rightarrow X_1 \geq_{icv} X_2$$

To prove the preceding theorem, we first prove the following proposition, which is of independent interest. It states that any two increasing functions of a random variable X have a nonnegative correlation.

Proposition 10.5.1   *If* $f(x)$ *and* $g(x)$ *are both increasing functions of* x, *then for any random variable* X

$$E[f(X)g(X)] \geq E[f(X)]E[g(X)]$$

If one of f *and* g *is an increasing function and the other is a decreasing function, then*
$$E[f(X)g(X)] \leq E[f(X)]E[g(X)]$$

Proof.   Let X and Y be independent with the same distribution, and suppose $f(x)$ and $g(x)$ are both increasing functions of x. Then $f(X) - f(Y)$ and $g(X) - g(Y)$ both have the same sign (both being nonnegative if $X \geq Y$ and being nonpositive if $X \leq Y$). Consequently,

$$(f(X) - f(Y))(g(X) - g(Y)) \geq 0$$

or, equivalently,

$$f(X)g(X) + f(Y)g(Y) \geq f(X)g(Y) + f(Y)g(X)$$

Taking expectations gives

$$E[f(X)g(X)] + E[f(Y)g(Y)] \geq E[f(X)g(Y)] + E[f(Y)g(X)]$$

Because X and Y are independent, the preceding yields

$$E[f(X)g(X)] + E[f(Y)g(Y)] \geq E[f(X)]E[g(Y)] \\ + E[f(Y)]E[g(X)]$$

Because X and Y have the same distribution, $E[f(Y)g(Y)] = E[f(X)g(X)]$ and $E[f(Y)] = E[f(X)]$, $E[g(Y)] = E[g(X)]$. Consequently, the preceding inequality yields

$$2E[f(X)g(X)] \geq 2E[f(X)]E[g(X)]$$

which is the desired result. Also, when f is decreasing and g is increasing, the preceding gives that

$$E[-f(X)g(X)] \geq E[-f(X)]E[g(X)]$$

Multiplying both sides by -1 now shows that

$$E[f(X)g(X)] \leq E[f(X)]E[g(X)]$$

which completes the proof. ☐

We will also need the following lemma.

Lemma 10.5.1 *If $E[X] = 0$ and $c \geq 1$ is a constant, then $X \geq_{icv} cX$.*

Proof. Let h be an increasing concave function, and let $c \geq 1$. The Taylor series expansion with remainder of $h(cx)$ about x gives that, for some w between x and cx,

$$\begin{aligned} h(cx) &= h(x) + h'(x)(cx - x) + h''(w)(cx - x)^2/2! \\ &\leq h(x) + h'(x)(cx - x) \end{aligned}$$

where the inequality follows because h concave implies that $h''(w) \leq 0$. Because the preceding holds for all x, it follows that

$$h(cX) \leq h(X) + (c - 1)Xh'(X)$$

Taking expectations gives

$$\begin{aligned} E[h(cX)] &\leq E[h(X)] + (c - 1)E[Xh'(X)] \\ &\leq E[h(X)] + (c - 1)E[X]E[h'(X)] \\ &= E[h(X)] \end{aligned}$$

where the second inequality follows from Proposition 10.5.1 because $f(x) = x$ is an increasing function and, because h is concave, $h'(x)$

is a decreasing function of x; and the final equality follows because $E[X] = 0$. □

We are now ready to prove Theorem 10.5.2.

Proof of Theorem 10.5.2 Assume that $\mu_1 \geq \mu_2$ and $\sigma_1 \leq \sigma_2$. Let Z be a normal random variable with mean 0 and variance 1. With $c = \sigma_2/\sigma_1 \geq 1$, it follows from Lemma 10.5.1 that $\sigma_1 Z \geq_{icv} c\sigma_1 Z = \sigma_2 Z$. Now, let $h(x)$ be a concave and increasing function of x. Then,

$$E[h(\mu_1 + \sigma_1 Z)] \geq E[h(\mu_2 + \sigma_1 Z)] \quad \text{(because } \mu_1 \geq \mu_2 \text{ and } h$$
$$\geq E[h(\mu_2 + \sigma_2 Z)] \quad \text{is increasing)}$$

where the final inequality follows because $g(x) = h(\mu_2 + x)$ is a concave, increasing function of x, and $\sigma_1 Z \geq_{icv} \sigma_2 Z$. The result now follows because $\mu_i + \sigma_i Z$ is a normal random variable with mean μ_i and variance σ_i^2. □

10.5.2 *More on Second-Order Dominance*

A useful result about second-order dominance is that if $X_1, \ldots, X_n$ and $Y_1, \ldots, Y_n$ are independent random vectors, then if X_i second-order stochastically dominates Y_i for each i, the sum of the X_i second-order stochastically dominates the sum of the Y_i.

Theorem 10.5.3 *Let* $X_1, \ldots, X_n$ *and* $Y_1, \ldots, Y_n$ *both be vectors of n independent random variables. If* $X_i \geq_{icv} Y_i$ *for each* $i = 1, \ldots, n$ *then* $\sum_{i=1}^{n} X_i \geq_{icv} \sum_{i=1}^{n} Y_i$.

Proof. Let h be an increasing concave function. We need to show that $E[h(\sum_{i=1}^{n} X_i)] \geq E[h(\sum_{i=1}^{n} Y_i)]$. The proof is by induction on n. Because the result is true when $n = 1$, assume it is true whenever the random vectors are of size $n - 1$. Now consider two vectors of independent random variables: $X_1, \ldots, X_n$ and $Y_1, \ldots, Y_n$. In addition suppose, without loss of generality, that these vectors are independent of each other. (It is "without loss of generality" because assuming that the two vectors are independent of each other does not affect the values of $E[h(\sum_{i=1}^{n} X_i)]$ and $E[h(\sum_{i=1}^{n} Y_i)]$, and thus a proof assuming vector independence is sufficient to prove the result.) To begin, we will show that $\sum_{i=1}^{n} X_i \geq_{icv} \sum_{i=1}^{n-1} Y_i + X_n$. To verify this, for any x define the function $h_x(a)$ by $h_x(a) = h(x+a)$ and note that h_x is an increasing concave function. Then, we have that

$$E\left[h\left(\sum_{i=1}^{n} X_i\right) \Big| X_n = x\right]$$

$$= E\left[h\left(x + \sum_{i=1}^{n-1} X_i\right) \Big| X_n = x\right]$$

$$= E\left[h\left(x + \sum_{i=1}^{n-1} X_i\right)\right] \quad \text{by independence}$$

$$= E\left[h_x\left(\sum_{i=1}^{n-1} X_i\right)\right]$$

$$\geq E\left[h_x\left(\sum_{i=1}^{n-1} Y_i\right)\right] \quad \text{by the induction hypothesis}$$

$$= E\left[h\left(x + \sum_{i=1}^{n-1} Y_i\right)\right]$$

$$= E\left[h\left(x + \sum_{i=1}^{n-1} Y_i\right) \Big| X_n = x\right] \quad \text{by independence}$$

$$= E\left[h\left(X_n + \sum_{i=1}^{n-1} Y_i\right) \Big| X_n = x\right]$$

Hence,

$$E\left[h\left(\sum_{i=1}^{n} X_i\right) \Big| X_n\right] \geq E\left[h\left(X_n + \sum_{i=1}^{n-1} Y_i\right) \Big| X_n\right]$$

and it follows, on taking expectations of the preceding, that

$$E\left[h\left(\sum_{i=1}^{n} X_i\right)\right] \geq E\left[h\left(X_n + \sum_{i=1}^{n-1} Y_i\right)\right]$$

Consequently, $\sum_{i=1}^{n} X_i \geq_{icv} \sum_{i=1}^{n-1} Y_i + X_n$. We now complete the proof by showing that $\sum_{i=1}^{n-1} Y_i + X_n \geq_{icv} \sum_{i=1}^{n} Y_i$. To do so, note that

$$E\left[h\left(\sum_{i=1}^{n-1} Y_i + X_n \right) \Big| \sum_{i=1}^{n-1} Y_i = y \right]$$

$$= E[h_y(X_n)] \geq E[h_y(Y_n)] = E\left[h\left(\sum_{i=1}^{n} Y_i \right) \Big| \sum_{i=1}^{n-1} Y_i = y \right]$$

where the inequality followed because h_y is an increasing, concave function and the equalities from the independence of the random variables. But the preceding gives that

$$E\left[h\left(\sum_{i=1}^{n-1} Y_i + X_n \right) \Big| \sum_{i=1}^{n-1} Y_i \right] \geq E\left[h\left(\sum_{i=1}^{n} Y_i \right) \Big| \sum_{i=1}^{n-1} Y_i \right]$$

Taking expectations of the preceding inequality yields that

$$E\left[h\left(\sum_{i=1}^{n-1} Y_i + X_n \right) \right] \geq E\left[h\left(\sum_{i=1}^{n} Y_i \right) \right]$$

Hence, $\sum_{i=1}^{n-1} Y_i + X_n \geq_{icv} \sum_{i=1}^{n} Y_i$, and the proof is complete. $\quad\square$

Remark. Theorem 10.5.3 along with the central limit theorem can be used to give another proof that a normal random variable decreases in second-order dominance as its variance increases. For suppose $\sigma_2 > \sigma_1$. Let X be equally likely to be plus or minus σ_1 and let Y be equally likely to be plus or minus σ_2. Then it is easy to directly verify that $X \geq_{icv} Y$ by showing that

$$h(-\sigma_1) + h(\sigma_1) \geq h(-\sigma_2) + h(\sigma_2)$$

whenever h is an increasing, concave function. (Because Y has the same distribution as $\frac{\sigma_2}{\sigma_1} X$, the result $X \geq_{icv} Y$ also follows from Lemma 10.5.1.) Now, let $X_i, i \geq 1$, be independent random variables all having the same distribution as X, and let $Y_i, i \geq 1$ be independent random variables all having the same distribution as Y. Then it follows from Theorem 10.5.3 that $\sum_{i=1}^{n} X_i \geq_{icn} \sum_{i=1}^{n} Y_i$. Because it is immediate that $W \geq_{icv} V$ implies that $cW \geq_{icv} cV$ for any positive constant c, we see that

$$\frac{\sum_{i=1}^{n} X_i}{\sqrt{n}} \geq_{icv} \frac{\sum_{i=1}^{n} Y_i}{\sqrt{n}}$$

The result now follows by letting $n \to \infty$, because the term on the left converges to a normal random variable with mean 0 and variance σ_1^2 and the term on the right converges to a normal random variable with mean 0 and variance σ_2^2. (Of course, to make this argument truly rigorous, we would need to show that second-order stochastic dominance is preserved when going to a limit.)

10.6   Exercises

Exercise 10.1 Suppose that

$$P(X_i = 1) = p_i = 1 - P(X_i = 0), \quad i = 1, 2$$

If $p_1 \geq p_2$, show that $X_1 \geq_{st} X_2$.

Exercise 10.2 Let $X(n, p)$ denote a binomial random variable with parameters n and p. Show that $X(n + 1, p) \geq_{st} X(n, p)$.

Exercise 10.3 Let $X(n, p)$ denote a binomial random variable with parameters n and p. If $p_1 \geq p_2$, show that $X(n, p_1) \geq_{st} X(n, p_2)$.

Exercise 10.4 If X_i is a normal random variable with mean μ_i and variance σ^2, for $i = 1, 2$, show that $X_1 \geq_{lr} X_2$ when $\mu_1 \geq \mu_2$.

Exercise 10.5 Let X_i be an exponential random variable with density function $f_i(x) = \lambda_i e^{-\lambda_i x}$, $i = 1, 2$. If $\lambda_1 \leq \lambda_2$, show that $X_1 \geq_{lr} X_2$.

Exercise 10.6 Let X_i be a Poisson random variable with mean λ_i. If $\lambda_1 \geq \lambda_2$, show that $X_1 \geq_{lr} X_2$.

Exercise 10.7 Show that $E[X] \geq_{icv} X$.

Exercise 10.8 Show that

$$h(-\sigma_1) + h(\sigma_1) \geq h(-\sigma_2) + h(\sigma_2)$$

whenever h is a concave function and $\sigma_2 > \sigma_1 > 0$.

Hint. Because h' is a decreasing function, $\int_{\sigma_1}^{\sigma_2} h'(x)\,dx \leq \int_{-\sigma_2}^{-\sigma_1} h'(x)\,dx$.

Exercise 10.9 If $X \geq_{icv} Y$, show that $g(X) \geq_{icv} g(Y)$ whenever g is an increasing concave function.

REFERENCES

[1] Ross, S., and E. Pekoz (2007). *A Second Course in Probability*, ProbabilityBookstore.com.

[2] Shaked, M., and J. G. Shanthikumar (1994). *Stochastic Orders and Their Applications*, Academic Press.

11. Optimization Models

11.1 Introduction

In this chapter we consider some optimization problems involving one-time investments not necessarily tied to the movement of a publicly traded security. Section 11.2 introduces a deterministic optimization problem where the objective is to determine an efficient algorithm for finding the optimal investment strategy when a fixed amount of money is to be invested in integral amounts among n projects, each having its own return function. Section 11.2.1 presents a dynamic programming algorithm that can always be used to solve the preceding problem; Section 11.2.2 gives a more efficient algorithm that can be employed when all the project return functions are concave; and Section 11.2.3 analyzes the special case, known as the knapsack problem, where project investments are made by purchasing integral numbers of shares, with each project return being a linear function of the number of shares purchased. Models in which probability is a key factor are considered in Section 11.3. Section 11.3.1 is concerned with a gambling model having an unknown win probability, and Section 11.3.2 examines a sequential investment allocation model where the number of investment opportunities is a random quantity.

11.2 A Deterministic Optimization Model

Suppose that you have m dollars to invest among n projects and that investing x in project i yields a (present value) return of $f_i(x)$, $i = 1, \ldots, n$. The problem is to determine the integer amounts to invest in each project so as to maximize the sum of the returns. That is, if we let x_i denote the amount to be invested in project i, then our problem (mathematically) is to

$$\text{choose nonnegative integers } x_1, \ldots, x_n$$
$$\text{such that } \sum_{i=1}^{n} x_i = m$$
$$\text{to maximize } \sum_{i=1}^{n} f_i(x_i).$$

11.2.1 A General Solution Technique Based on Dynamic Programming

To solve the preceding problem, let $V_j(x)$ denote the maximal possible sum of returns when we have a total of x to invest in projects $1, \ldots, j$. With this notation, $V_n(m)$ represents the maximal value of the problem posed in Section 11.2. Our determination of $V_n(m)$, and of the optimal investment amounts begins by finding the values of $V_j(x)$ for $x = 1, \ldots, m$, first for $j = 1$, then for $j = 2$, and so on up to $j = n$.

Because the maximal return when x must be invested in project 1 is $f_1(x)$, we have that

$$V_1(x) = f_1(x).$$

Now suppose that x must be invested between projects 1 and 2. If we invest y in project 2 then a total of $x - y$ is available to invest in project 1. Because the best return from having $x - y$ available to invest in project 1 is $V_1(x - y)$, it follows that the maximal sum of returns possible when the amount y is invested in project 2 is $f_2(y) + V_1(x - y)$. As the maximal sum of returns possible is obtained by maximizing the preceding over y, we see that

$$V_2(x) = \max_{0 \le y \le x} \{ f_2(y) + V_1(x - y) \}.$$

In general, suppose that x must be invested among projects $1, \ldots, j$. If we invest y in project j then a total of $x - y$ is available to invest in projects $1, \ldots, j-1$. Because the best return from having $x - y$ available to invest in projects $1, \ldots, j - 1$ is $V_{j-1}(x - y)$, it follows that the maximal sum of returns possible when the amount y is invested in project j is $f_j(y) + V_{j-1}(x - y)$. As the maximal sum of returns possible is obtained by maximizing the preceding over y, we see that

$$V_j(x) = \max_{0 \le y \le x} \{ f_j(y) + V_{j-1}(x - y) \}.$$

If we let $y_j(x)$ denote the value (or a value if there is more than one) of y that maximizes the right side of the preceding equation, then $y_j(x)$ is the optimal amount to invest in project j when you have x to invest among projects $1, \ldots, j$.

The value of $V_n(m)$ can now be obtained by first determining $V_1(x)$, then $V_2(x), V_3(x), \ldots, V_{n-1}(x)$ and finally $V_n(m)$. The optimal amount to invest in project n would be given by $y_n(m)$; the optimal amount to invest in project $n-1$ would be $y_{n-1}(m - y_n(m))$, and so on.

This solution approach – which views the problem as involving n sequential decisions and then analyzes it by determining the optimal last decision, then the optimal next to last decision, and so on – is called *dynamic programming*. (Dynamic programming was previously used in Section 8.3 for pricing, and finding, the optimal exercise strategy for an American put option.)

Example 11.2a Suppose that three investment projects with the following return functions are available:

$$f_1(x) = \frac{10x}{1+x}, \quad x = 0, 1, \ldots,$$

$$f_2(x) = \sqrt{x}, \quad x = 0, 1, \ldots,$$

$$f_3(x) = 10(1 - e^{-x}), \quad x = 0, 1, \ldots,$$

and that we want to maximize our return when we have 5 to invest. Now,

$$V_1(x) = f_1(x) = \frac{10x}{1+x}, \quad y_1(x) = x.$$

Because

$$V_2(x) = \max_{0 \le y \le x} \{f_2(y) + V_1(x - y)\} = \max_{0 \le y \le x} \left\{ \sqrt{y} + \frac{10(x - y)}{1 + x - y} \right\},$$

we see that

$$V_2(1) = \max\{10/2, 1\} = 5, \quad y_2(1) = 0,$$

$$V_2(2) = \max\{20/3, 1 + 5, \sqrt{2}\} = 20/3, \quad y_2(2) = 0,$$

$$V_2(3) = \max\{30/4, 1 + 20/3, \sqrt{2} + 5, \sqrt{3}\} = 23/3, \quad y_2(3) = 1,$$

$$V_2(4) = \max\{40/5, 1 + 30/4, \sqrt{2} + 20/3, \sqrt{3} + 5, \sqrt{4}\}$$
$$= 8.5, \quad y_2(4) = 1,$$

$$V_2(5) = \max\{50/6, 1 + 8, \sqrt{2} + 7.5, \sqrt{3} + 20/3, \sqrt{4} + 5, \sqrt{5}\}$$
$$= 9, \quad y_2(5) = 1.$$

Continuing, we have that

$$V_3(x) = \max_{0 \le y \le x} \{f_3(y) + V_2(x - y)\} = \max_{0 \le y \le x} \{10(1 - e^{-y}) + V_2(x - y)\}.$$

Using that

$$1 - e^{-1} = .632, \quad 1 - e^{-2} = .865, \quad 1 - e^{-3} = .950,$$
$$1 - e^{-4} = .982, \quad 1 - e^{-5} = .993,$$

we obtain

$$V_3(5) = \max\{9, \ 6.32 + 8.5, \ 8.65 + 23/3,$$
$$9.50 + 20/3, \ 9.82 + 5, \ 9.93\} = 16.32,$$
$$y_3(5) = 2.$$

Thus, the maximal sum of returns from investing 5 is 16.32; the optimal amount to invest in project 3 is $y_3(5) = 2$; the optimal amount to invest in project 2 is $y_2(3) = 1$; and the optimal amount to invest in project 1 is $y_1(2) = 2$. □

11.2.2   A Solution Technique for Concave Return Functions

More efficient algorithms for solving the preceding problem are available when the return functions satisfy certain conditions. For instance, suppose that each of the functions $f_i(x)$ is concave, where a function $g(i)$, $i = 0, 1, \ldots$, is said to be concave if

$$g(i + 1) - g(i) \text{ is nonincreasing in } i.$$

That is, a return function would be concave if the additional (or marginal) gain from each additional unit invested becomes smaller as more has already been invested.

Let us now assume that the functions $f_i(x)$, $i = 1, \ldots, n$, are all concave, and again consider the problem of choosing nonnegative integers $x_1, \ldots, x_n$, whose sum is m, to maximize $\sum_{i=1}^{n} f_i(x_i)$. Suppose that $x_1^o, \ldots, x_n^o$ is an optimal vector for this problem: a vector of nonnegative integers that sum to m and with

$$\sum_{i=1}^{n} f_i(x_i^o) = \max \sum_{i=1}^{n} f_i(x_i),$$

where the maximum is over all nonnegative integers $x_1, \ldots, x_n$ that sum to m. Now suppose that we have a total of $m + 1$ to invest. We will argue that there is an optimal vector $y_1^o, \ldots, y_n^o$ with $\sum_{i=1}^n y_i^o = m + 1$ that satisfies

$$y_i^o \geq x_i^o, \quad i = 1, \ldots, n. \tag{11.1}$$

To verify (11.1), suppose we have $m + 1$ to invest and consider any investment strategy $y_1, \ldots, y_n$ with $\sum_{i=1}^n y_i = m + 1$ such that, for some value of k,

$$y_k < x_k^o.$$

Because $m + 1 = \sum_i y_i > \sum_i x_i^o = m$, it follows that there must be a j such that

$$x_j^o < y_j.$$

We will now argue that when you have $m + 1$ to invest, the investment strategy that invests $y_k + 1$ in project k, $y_j - 1$ in project j, and y_i in project i for $i \neq k$ or j is at least as good as the strategy that invests y_i in project i for each i. To verify that this new investment strategy is at least as good as the original y-strategy, we need to show that

$$f_k(y_k + 1) + f_j(y_j - 1) \geq f_k(y_k) + f_j(y_j)$$

or, equivalently, that

$$f_k(y_k + 1) - f_k(y_k) \geq f_j(y_j) - f_j(y_j - 1). \tag{11.2}$$

Because $x_1^o, \ldots, x_n^o$ is optimal when there is m to invest, it follows that

$$f_k(x_k^o) + f_j(x_j^o) \geq f_k(x_k^o - 1) + f_j(x_j^o + 1)$$

or, equivalently, that

$$f_k(x_k^o) - f_k(x_k^o - 1) \geq f_j(x_j^o + 1) - f_j(x_j^o). \tag{11.3}$$

Consequently,

$$
\begin{aligned}
f_k(y_k + 1) &- f_k(y_k) \\
&\geq f_k(x_k^o) - f_k(x_k^o - 1) \quad \text{(by concavity, since } y_k + 1 \leq x_k^o) \\
&\geq f_j(x_j^o + 1) - f_j(x_j^o) \quad \text{(by (11.3))} \\
&\geq f_j(y_j) - f_j(y_j - 1) \quad \text{(by concavity, since } x_j^o + 1 \leq y_j).
\end{aligned}
$$

Thus, we have verified the inequality (11.2), which shows that any strategy for investing $m + 1$ that calls for investing less than x_k^o in some project k can be at least matched by one whose investment in project k is increased by 1 with a corresponding decrease in some project j whose investment was greater than x_j^o. Repeating this argument shows that, for any strategy of investing $m + 1$, we can find another strategy that invests at least x_i^o in project i for all $i = 1, \ldots, n$ and yields a return that is at least as large as the original strategy. But this implies that we can find an optimal strategy $y_1^o, \ldots, y_n^o$ for investing $m + 1$ that satisfies the inequality (11.1).

Because the optimal strategy for investing $m + 1$ invests at least as much in each project as does the optimal strategy for investing m, it follows that the optimal strategy for $m + 1$ can be found by using the optimal strategy for m and then investing the extra dollar in that project whose marginal increase is largest. Therefore, we can find the optimal investment (when we have m) by first solving the optimal investment problem when we have 1 to invest, then when we have 2, then 3, and so on.

Example 11.2b Let us reconsider Example 11.2a, where we have 5 to invest among three projects whose return functions are

$$f_1(x) = \frac{10x}{1 + x},$$

$$f_2(x) = \sqrt{x},$$

$$f_3(x) = 10(1 - e^{-x}).$$

Let $x_i(j)$ denote the optimal amount to invest in project i when we have a total of j to invest. Because

$$\max\{f_1(1), f_2(1), f_3(1)\} = \max\{5, 1, 6.32\} = 6.32,$$

we see that

$$x_1(1) = 0, \quad x_2(1) = 0, \quad x_3(1) = 1.$$

Since

$$\max_i\{f_i(x_i(1) + 1) - f_i(x_i(1))\} = \max\{5, 1, 8.65 - 6.32\} = 5,$$

we have

$$x_1(2) = 1, \quad x_2(2) = 0, \quad x_3(2) = 1.$$

Because

$$\max_i \{f_i(x_i(2) + 1) - f_i(x_i(2))\} = \max\{20/3 - 5, 1, 8.65 - 6.32\}$$
$$= 2.33,$$

it follows that

$$x_1(3) = 1, \quad x_2(3) = 0, \quad x_3(3) = 2.$$

Since

$$\max_i \{f_i(x_i(3) + 1) - f_i(x_i(3))\} = \max\{20/3 - 5, 1, 9.50 - 8.65\}$$
$$= 1.67,$$

we obtain

$$x_1(4) = 2, \quad x_2(4) = 0, \quad x_3(4) = 2.$$

Finally,

$$\max_i \{f_i(x_i(4) + 1) - f_i(x_i(4))\} = \max\{30/4 - 20/3, 1, 9.50 - 8.65\}$$
$$= 1,$$

giving that

$$x_1(5) = 2, \quad x_2(5) = 1, \quad x_3(5) = 2.$$

The maximal return is thus $6.32 + 5 + 2.33 + 1.67 + 1 = 16.32$.   □

The following algorithm can be used to solve the problem when m is to be invested among n projects, each of which has a concave return function. The quantity k will represent the current amount to be invested, and x_i will represent the optimal amount to invest in project i when a total of k is to be invested.

Algorithm

(1) Set $k = 0$ and $x_i = 0$, $i = 1, \ldots, n$.
(2) $m_i = f_i(x_i + 1) - f_i(x_i)$, $i = 1, \ldots, n$.
(3) $k = k + 1$.
(4) Let J be such that $m_J = \max_i m_i$.

(5) If $J = j$, then

$$x_j \rightarrow x_j + 1,$$

$$m_j \rightarrow f_j(x_j + 1) - f_j(x_j).$$

(6) If $k < m$, go to step (3).

Step (5) means that if the value of J is j, then (a) the value of x_j should be increased by 1 and (b) the value of m_j should be reset to equal the difference of f_j evaluated at 1 plus the new value of x_j and f_j evaluated at the new value of x_j.

Remark: When $g(x)$ is defined for all x in an interval, then g is concave if $g'(t)$ is a decreasing function of t (that is, if $g''(t) \leq 0$). Hence, for g concave

$$\int_i^{i+1} g'(s)ds \leq \int_{i-1}^i g'(s)ds$$

yielding that

$$g(i+1) - g(i) \leq g(i) - g(i-1)$$

which we used as the definition of concavity for g defined on the integers.

11.2.3 The Knapsack Problem

Suppose one invests in project i by buying an integral number of shares in that project, with each share costing c_i and returning v_i. If we let x_i denote the number of shares of project i that are purchased, then the problem – when one can invest at most m in the n projects – is to

choose nonnegative integers $x_1, \ldots, x_n$

such that $\sum_{i=1}^n x_i c_i \leq m$

to maximize $\sum_{i=1}^n v_i x_i$.

We will use a dynamic programming approach to solve this problem. To begin, let $V(x)$ be the maximal return possible when we have x to invest. If we start by buying one share of project i, then a return v_i will be received and we will be left with a capital of $x - c_i$. Because $V(x - c_i)$

is the maximal return that can be obtained fom the amount $x - c_i$, it follows that the maximal return possible if we have x and begin investing by buying one share of project i is

maximal return if start by purchasing one share of $i = v_i + V(x - c_i)$.

Hence $V(x)$, the maximal return that can be obtained from the investment capital x, satisfies

$$V(x) = \max_{i:c_i \leq x}\{v_i + V(x - c_i)\}. \tag{11.4}$$

Let $i(x)$ denote the value of i that maximizes the right side of (11.4). Then, when one has x, it is optimal to purchase one share of project $i(x)$. Starting with

$$V(1) = \max_{i:c_i \leq 1} v_i,$$

it is easy to determine the values of $V(1)$ and $i(1)$, which will then enable us to use Equation (11.4) to determine $V(2)$ and $i(2)$, and so on.

Remark. This problem is called a *knapsack* problem because it is mathematically equivalent to determining the set of items to be put in a knapsack that can carry a total weight of at most m when there are n different types of items, with each type i item having weight c_i and yielding the value v_i.

Example 11.2c Suppose you have 25 to invest among three projects whose cost and return values are as follows.

Project	Cost per share	Return per share
1	5	7
2	9	12
3	15	22

Then

$$V(x) = 0, \quad x \le 4,$$

$$V(x) = 7, \quad i(x) = 1, \quad x = 5, 6, 7, 8,$$

$$V(9) = \max\{7 + V(4), 12 + V(0)\} = 12, \quad i(9) = 2,$$

$$V(x) = \max\{7 + V(x - 5), 12 + V(x - 9)\} = 14,$$

$$i(x) = 1, \quad x = 10, 11, 12, 13,$$

$$V(14) = \max\{7 + V(9), 12 + V(5)\} = 19, \quad i(x) = 1 \text{ or } 2,$$

$$V(15) = \max\{7 + V(10), 12 + V(6), 22 + V(0)\} = 22, \quad i(15) = 3,$$

$$V(16) = \max\{7 + V(11), 12 + V(7), 22 + V(1)\} = 22, \quad i(16) = 3,$$

$$V(17) = \max\{7 + V(12), 12 + V(8), 22 + V(2)\} = 22, \quad i(17) = 3,$$

$$V(18) = \max\{7 + V(13), 12 + V(9), 22 + V(3)\} = 24, \quad i(18) = 2,$$

and so on. Thus, for instance, with 18 it is optimal to first purchase one share of project $i(18) = 2$ and then purchase one share of project $i(9) = 2$. That is, with 18 it is optimal to purchase two shares of project 2 for a total return of 24. □

11.3 Probabilistic Optimization Problems

In this section we consider two optimization problems that are probabilistic in nature. Section 11.3.1 deals with a gambling model that has been chosen to illustrate the value of information. Section 11.3.2 is concerned with an investment allocation problem when the number of investment opportunities is random.

11.3.1 *A Gambling Model with Unknown Win Probabilities*

Suppose, in Example 9.2a, that an investment's win probability p is not fixed but can be one of three possible values: $p_1 = .45$, $p_2 = .55$, or $p_3 = .65$. Suppose also that it will be p_1 with probability $1/4$, p_2 with probability $1/2$, and p_3 with probability $1/4$. If an investor does not have information about which p_i has been chosen, then she will take the win probability to be

$$p = \tfrac{1}{4}p_1 + \tfrac{1}{2}p_2 + \tfrac{1}{4}p_3 = .55.$$

Assuming (as in Example 9.2a) a log utility function, it follows from the results of that example that the investor will invest $100(2p - 1) = 10\%$ of her fortune, with the expected utility of her final fortune being

$$\log(x) + .55\log(1.1) + .45\log(.9) = \log(x) + .0050 = \log(e^{.0050}x),$$

where x is the investor's initial fortune.

Suppose now that the investor is able to learn, before making her investment, which p_i is the win probability. If .45 is the win probability, then the investor will not invest and so the conditional expected utility of her final fortune will be $\log(x)$. If .55 is the win probability, the investor will do as shown previously, and the conditional expected utility of her final fortune will be $\log(x) + .0050$. Finally, if .65 is the win probability, the investor will invest 30% of her fortune and the conditional expected utility of her final fortune will be

$$\log(x) + .65\log(1.3) + .35\log(.7) = \log(x) + .0456.$$

Therefore, the expected final utility of an investor who will learn which p_i is the win probability before making her investment is

$$\frac{1}{4}\log(x) + \frac{1}{2}(\log(x) + .0050) + \frac{1}{4}(\log(x) + .0456) = \log(x) + .0139$$
$$= \log(e^{.0139}x).$$

11.3.2   *An Investment Allocation Model*

An investor has the amount D available to invest. During each of N time instants, an opportunity to invest will (independently) present itself with probability p. If the opportunity occurs, the investor must decide how much of her remaining wealth to invest. If y is invested in an opportunity then $R(y)$, a specified function of y, is earned at the end of the problem. Assuming that both the amount invested and the return from that investment become unavailable for future investment, the problem is to determine how much to invest at each opportunity so as to maximize the expected value of the investor's final wealth, which is equal to the sum of all the investment returns and the amount that was never invested.

To solve this problem, let $W_n(x)$ denote the maximal expected final wealth when the investor has x to invest and there are n time instants in the problem; let $V_n(x)$ denote the maximal expected final wealth when the investor has x to invest, there are n time instants in the problem, and an opportunity is at hand. To determine an equation for $V_n(x)$, note that if y is initially invested then the investor's maximal expected final wealth will be $R(y)$ plus the maximal expected amount that she can obtain in $n - 1$ time instants when her investment capital is $x - y$. Because this latter quantity is $W_{n-1}(x - y)$, we see that the maximal expected final wealth when y is invested is $R(y) + W_{n-1}(x - y)$. The investor can now choose y to maximize this sum, so we obtain the equation

$$V_n(x) = \max_{0 \le y \le x} \{R(y) + W_{n-1}(x - y)\}. \qquad (11.5)$$

When the investor has x to invest and there are n time instants to go, either an opportunity occurs and the maximal expected final wealth is $V_n(x)$, or an opportunity does not occur and the maximal expected final wealth is $W_{n-1}(x)$. Because each opportunity occurs with probability p, it follows that

$$W_n(x) = pV_n(x) + (1 - p)W_{n-1}(x). \qquad (11.6)$$

Starting with $W_0(x) = x$, we can use Equation (11.5) to obtain $V_1(x)$ for all $0 \le x \le D$, then use Equation (11.6) to obtain $W_1(x)$ for all $0 \le x \le D$, then use Equation (11.5) to obtain $V_2(x)$ for all $0 \le x \le D$, then use Equation (11.6) to obtain $W_2(x)$, and so on. If we let $y_n(x)$ be the value of y that maximizes the right side of Equation (11.5), then the optimal policy is to invest the amount $y_n(x)$ if there are n time instants remaining, an opportunity is present, and our current investment capital is x.

Example 11.3a Suppose that we have 10 to invest, there are two time instants, an opportunity will present itself each instant with probability $p = .7$, and

$$R(y) = y + 10\sqrt{y}.$$

Find the maximal expected final wealth as well as the optimal policy.

Solution. Starting with $W_0(x) = x$, Equation (11.5) gives

$$V_1(x) = \max_{0 \le y \le x} \{y + 10\sqrt{y} + x - y\}$$

$$= x + \max_{0 \le y \le x} \{10\sqrt{y}\}$$

$$= x + 10\sqrt{x}$$

and $y_1(x) = x$. Thus,

$$W_1(x) = .7(x + 10\sqrt{x}) + .3x = x + 7\sqrt{x},$$

yielding that

$$V_2(x) = \max_{0 \le y \le x} \{y + 10\sqrt{y} + x - y + 7\sqrt{x - y}\}$$

$$= x + \max_{0 \le y \le x} \{10\sqrt{y} + 7\sqrt{x - y}\}$$

$$= x + \sqrt{149x}, \tag{11.7}$$

where calculus gave the final equation as well as the result:

$$y_2(x) = \frac{100}{149}x. \tag{11.8}$$

The preceding now yields

$$W_2(x) = .7(x + \sqrt{149x}) + .3(x + 7\sqrt{x}) = x + .7\sqrt{149x} + 2.1\sqrt{x}.$$

Thus, starting with 10, the maximal expected final wealth is

$$W_2(10) = 10 + .7\sqrt{1490} + 2.1\sqrt{10} = 43.66.$$

Hence the optimal policy is to invest $\frac{1000}{149} = 6.71$ if an opportunity presents itself at the initial time instant and then to invest whatever of your fortune remains if an opportunity presents itself at the final time instant. □

Provided that $R(y)$ is a nondecreasing concave function, the following result can be proved.

Theorem 11.3.1 *If $R(y)$ is a nondecreasing concave function, then:*

(a) *$V_n(x)$ and $W_n(x)$ are both nondecreasing concave functions;*

(b) *$y_n(x)$ is a nondecreasing function of x;*

(c) $x - y_n(x)$ *is a nondecreasing function of* x; *and*

(d) $y_n(x)$ *is a nonincreasing function of* n.

Parts (b) and (c) state, respectively, that the more you have the more you should invest and that the more you have the more you should conserve. Part (d) says that the more time you have the less you should invest each time.

11.4   Exercises

Exercise 11.1   Find the optimal investment strategy when 6 is to be invested between two projects having return functions

$$f_1(x) = 2\log(x+1), \quad f_2(x) = \sqrt{x}, \quad x = 0, 1, \dots.$$

Exercise 11.2   Find the optimal strategy and the maximal return in Example 11.2a when you have 8 to invest. Use the method of Example 11.2a.

Exercise 11.3   Use the method of Example 11.2b to solve the preceding exercise.

Exercise 11.4   The function $g(i)$, $i = 0, 1, \dots$, is said to be *convex* if

$$g(i+1) - g(i) \text{ is nondecreasing in } i.$$

Show that, if all return functions are convex, then there is an optimal investment strategy for the problem of Section 11.2 that invests everything in a single project.

Exercise 11.5   Consider the problem of choosing nonnegative integers $x_1, \dots, x_n$, whose sum is $m = kn$, to maximize

$$f(x_1, \dots, x_n) = \sum_{i=1}^{n} f(x_i),$$

where $f(x)$ is a specified function for which $f(0) = 0$.

(a) If $f(x)$ is concave, show that the maximal value is $nf(k)$.

(b) If $f(x)$ is convex, show that the maximal value is $f(kn)$.

Exercise 11.6   Continue with Example 11.2c and find the optimal strategy when you have 25 to invest.

Exercise 11.7   Starting with some initial wealth, you must decide in each of the following N periods how much of your wealth to invest and how much to consume. Assume the utility that you attain from consuming the amount x during a period is $\sqrt{x}$ and that your objective is to maximize the sum of the utilities you obtain in the N periods. Assume also that an investment earns a fixed rate of return r per period. Let $V_n(x)$ denote the maximal sum of utilities that can be attained when one's current fortune is x and n additional periods remain.

(a) What is the value of $V_1(x)$?
(b) Find $V_2(x)$.
(c) Derive an equation for $V_n(x)$.
(d) Determine the optimal amounts to invest and to consume when your fortune is x and you have n periods remaining.

Hint.   Let the decision be the fraction of your wealth to consume.

Exercise 11.8   An individual begins processing n jobs at time 0. Job i takes time x_i to process. If the processing of job i is completed at time t, then the processor earns the return $R_i(t)$. Jobs may be processed in any order, with the objective being to maximize the sum of the processor's returns. For any subset S of jobs, let $V(S)$ be the maximal return that the processor can receive from the jobs in S when all the jobs not in S have already been processed. For instance, $V(\{1, 2, \ldots, n\})$ is the maximal return that can be earned.

(a) Derive an equation that relates $V(S)$ to V evaluated at different subsets of S.
(b) Explain how the result of part (a) can be used to find the optimal policy.

Exercise 11.9   An investor must choose between one of two possible investments. In the first investment, she must choose an amount to be at risk, and she will then either win that amount with probability .6 or lose it with probability .4. In the second investment, there is a 70-percent chance that the win probability will be .4 and a 30-percent chance that it will be .8. Although the investor must decide on the investment project before she learns the win probability for the second investment, if she chooses that investment then she will be told the win probability *before* she chooses the amount to risk. Which investment should she choose and how much should she risk if she has a logarithmic utility function?

Exercise 11.10 Verify Equations (11.7) and (11.8).

Exercise 11.11 Consider a graph with nodes $1, \ldots, m$ and edges (i, j), $i \neq j$. Suppose that the time it takes to traverse the edge (i, j) depends on when one begins traveling along that edge. Specifically, suppose the time is $t_s(i, j)$ if one leaves node i at time s. For specified nodes 1 and m, the problem of interest is to find the path from node 1 to node m that minimizes the time at which node m is reached when one begins at node 1 at time 0. For instance, if the path $1, i_1, \ldots, i_k = m$ is used, the the time at which node m is reached is $a_1 + \ldots + a_k$, where

$$a_1 = t_0(1, i_1)$$
$$a_2 = t_{a_1}(i_1, i_2)$$
$$a_3 = t_{a_1+a_2}(i_2, i_3)$$
$$a_k = t_{a_1+\ldots+a_{k-1}}(i_{k-1}, i_k)$$

Let $T(j)$ denote the minimal time that node j can be reached if one starts at node 1 at time 0. Argue that

$$T(j) = \min_i \{T(i) + t_{T(i)}(i, j)\}$$

Assume that $s + t_s(i, j)$ increases in s. That is, if one reaches node i at time s and then goes directly to node j then the time to arrive at node j increases in s.

12. Stochastic Dynamic Programming

12.1 The Stochastic Dynamic Programming Problem

In the general stochastic dynamic programming problem, we suppose that a system is observed at the beginning of each period and its state is determined. Let S denote the set of all possible states. After observing the state of the system, an action must be chosen. If the state is x and action a is chosen, then

(a) a reward $r(x, a)$ is earned; and
(b) the next state, call it $Y(x, a)$, is a random variable whose distribution depends only on x and a.

Suppose our objective is to maximize the expected sum of rewards that can be earned over N time periods. To attack this problem, let $V_n(x)$ denote the maximal expected sum of rewards that can be earned in the next n time periods given that the current state is x. Now, if we initially choose action a, then a reward $r(x, a)$ is immediately earned, and the next state will be $Y(x, a)$. If $Y(x, a) = y$, then at that point there will be an additional $n - 1$ time periods to go, and so the maximal expected additional return we could earn from then on would be $V_{n-1}(y)$. Hence, if the current state is x, then the maximal expected return that could be earned over the next n time periods *if we initially choose action a* is

$$r(x, a) + E[V_{n-1}(Y(x, a))]$$

Hence, $V_n(x)$, the overall maximal expected return, satisfies

$$V_n(x) = \max_a \{ r(x, a) + E[V_{n-1}(Y(x, a))] \} \qquad (12.1)$$

Starting with $V_0(x) = 0$ the preceding equation can be used to recursively solve for the functions $V_1(x)$, then $V_2(x)$, and on up to $V_N(x)$. The policy that, when there are n additional time periods to go with the current state being x, chooses the action (or one of the actions) that maximizes the right side of the preceding is an optimal policy. That is, if we let $a_n(x)$ equal the action that maximizes $r(x, a) + E[V_{n-1}(Y(x, a))]$, written as

$$a_n(x) = \arg\max_a \{r(x, a) + E[V_{n-1}(Y(x, a))]\}, \quad n = 1, \ldots, N$$

then the policy that, for all n and x, chooses action $a_n(x)$ when the state is x and there are are n time periods remaining is an optimal policy.

The function $V_n(x)$ is called the *optimal value function*, and Equation (12.1) is called the *optimality equation*.

When S is a subset of the set of all integers, we let $P_{i,a}(j)$ denote the probability that the next state is j when the current state is i and action a is chosen. In this case, the optimality equation can be written

$$V_n(i) = \max_a \left\{ r(i, a) + \sum_j P_{i,a}(j) V_{n-1}(j) \right\}$$

When S is a continuous set, we let $f_{x,a}(y)$ be the probability density of the next state given that the current state is x and action a is chosen. In this case, the optimality equation can be written

$$V_n(x) = \max_a \left\{ r(x, a) + \int f_{x,a}(y) V_{n-1}(y) dy \right\}$$

In certain problems future costs may be discounted. Specifically, a cost incurred k time periods in the future may be discounted by the factor β^k. In such cases the optimality equation becomes

$$V_n(x) = \max_a \{r(x, a) + \beta E[V_{n-1}(Y(x, a))]\}$$

For instance, if we wanted to maximize the present value of the sum of rewards, then we would let $\beta = \frac{1}{1+r}$, where r is the interest rate per period. The quantity β is called the *discount factor* and is usually assumed to satisfy $0 \le \beta \le 1$.

Example 12.1a *Optimal Return from a Call Option*

Suppose the following discrete time model for the price movement of a security: whatever the price history so far, the price of the security during the following period is its current price multiplied by a random variable Y. Assume an interest rate of $r > 0$ per period, let $\beta = \frac{1}{1+r}$, and suppose that we want to determine the appropriate value of an American call option having exercise K and expiring at the end of n additional periods. Because we are not assuming that Y has only two possible values, there will not be a unique risk-neutral probability law, and so arbitrage considerations will not enable us to determine the value of the option. Moreover, because we shall suppose that the security cannot be sold short for the market price, there will no longer be an arbitrage argument against early exercising. To determine the appropriate value of the option under these conditions, we will suppose that the successive Y's are independent with a common specified distribution, and take as our objective the determination of the maximal expected present-value return that can be obtained from the option.

As the dynamic programming state of the system will be the current price, let us define $V_j(x)$, the optimal value function, to equal the maximal expected present-value return from the option given that it has not yet been exercised, a total of j periods remain before the option expires, and the current price of the security is x. Now, if the preceding is the situation and the option is exercised, then a return $x - K$ is earned and the problem ends; on the other hand, if the option is not exercised, then the maximal expected present-value return will be $E[\beta V_{j-1}(xY)]$. Because the overall best is the maximum of the best one can obtain under the different possible actions, we see that the optimality equation is

$$V_j(x) = \max\{x - K, \quad \beta E[V_{j-1}(xY)]\}$$

with the boundary condition

$$V_0(x) = (x - K)^+ = \max\{x - K, \quad 0\}$$

The policy that, when the current price is x and j periods remain before the option expires, exercises if $V_j(x) = x - K$ and does not exercise if $V_j(x) > x - K$ is an optimal policy. (That is, the optimal policy exercises in state x when j periods remain if and only if $V_j(x) = x - K$.)

We now determine the structure of the optimal policy. Specifically, we show that if $E[Y] \geq 1 + r$, then the call option should never be exercised early; whereas if $E[Y] < 1 + r$, then there is a nondecreasing sequence x_j, $j \geq 0$, such that the policy that exercises when j periods remain if the current price is at least x_j is an optimal policy. To establish the preceding, we will need some preliminary results.

Lemma 12.1.1 *If $E[Y] \geq 1 + r$, then the policy that only exercises when no additional time remains and the price is greater than K is an optimal policy.*

Proof. It follows from the optimality equation that $V_j(x) \geq x - K$. Using that $\beta E[Y] \geq \beta(1 + r) = 1$, we see that, for $j \geq 1$,

$$\beta E[V_{j-1}(xY)] \geq \beta E[xY - K] \geq x - \beta K > x - K$$

Thus, it is never optimal to exercise early. $\square$

Lemma 12.1.2 *If $E[Y] < 1 + r$, then $V_j(x) - x$ is a decreasing function of x.*

Proof. The proof is by induction on j. Because

$$V_0(x) - x = \max\{-K, \quad -x\}$$

the result is true when $j = 0$. So, assume that $V_{j-1}(x) - x$ is decreasing in x. Then, by the optimality equation,

$$
\begin{aligned}
V_j(x) - x &= \max\{-K, \quad \beta E[V_{j-1}(xY)] - x\} \\
&= \max\{-K, \quad \beta(E[V_{j-1}(xY) - xE[Y]]) + \beta x E[Y] - x\} \\
&= \max\{-K, \quad \beta E[V_{j-1}(xY) - xY] + x(\beta E[Y] - 1)\}
\end{aligned}
$$

Now, by the induction hypothesis, for any value of Y, $V_{j-1}(xY) - xY$ is decreasing in x, and therefore so is $E[V_{j-1}(xY) - xY]$. Because $\beta E[Y] < 1$, it also follows that $x(\beta E[Y] - 1)$ is decreasing in x. Hence, $\beta E[V_{j-1}(xY) - xY] + x(\beta E[Y] - 1)$, and thus $V_j(x) - x$, is decreasing in x, which completes the proof. $\square$

Proposition 12.1.1 *If $E[Y] < 1 + r$, then there is a increasing sequence x_j, $j \geq 0$ such that the policy that exercises when j periods remain whenever the current price is at least x_j is an optimal policy.*

Proof. Let $x_j = \min\{x : V_j(x) = x - K\}$ be the minimal price at which it is optimal to exercise when j periods remain. It follows from Lemma 12.1.2 that for $x' > x_j$,

$$V_j(x') - x' \leq V_j(x_j) - x_j = -K$$

Because the optimality equation yields that $V_j(x') \geq x' - K$, we see that

$$V_j(x') = x' - K$$

thus showing that it is optimal to exercise when j stages remain and the current price is x' if and only if $x' \geq x_j$. To show that x_j increases in j, we use that $V_j(x)$ is increasing in j, which follows because having additional time before the option expires cannot reduce the maximal expected return. Using this yields that

$$V_{j-1}(x_j) \leq V_j(x_j) = x_j - K$$

Because the optimality equation yields that $V_{j-1}(x_j) \geq x_j - K$, the preceding equation shows that

$$V_{j-1}(x_j) = x_j - K$$

Because x_{j-1} is defined as the smallest value of x for which $V_{j-1}(x) = x - K$, the preceding yields that $x_{j-1} \leq x_j$ and completes the proof. $\square$

Although we have assumed that $r(x, a)$, the reward earned when action a is chosen in state x, is a constant, it sometimes is the case the reward is a random variable that is independent of all that has previously occurred. In such cases $r(x, a)$ should be interpreted as the expected reward earned.

Example 12.1b An urn initially has n red and m blue balls. At each stage the player may randomly choose a ball from the urn; if the ball is red, then 1 is earned, and if it is blue, then 1 is lost. The chosen ball is discarded. At any time the player can decide to stop playing. To maximize the player's total expected net return, we analyze this as a dynamic programming problem with the state equal to the current composition of the urn. We let $V(r, b)$ denote the maximum expected additional return given that there are currently r red and b blue balls in the urn. Now, the expected immediate reward if a ball is chosen in state (r, b) is $\frac{r}{r+b} - \frac{b}{r+b} = \frac{r-b}{r+b}$. Because the best one can do after the initial draw is $V(r-1, b)$ if a red ball is chosen, or $V(r, b-1)$ if a blue ball is chosen, we see that the optimality equation is

$$V(r, b) = \max\left\{0, \; \frac{r-b}{r+b} + \frac{r}{r+b}V(r-1, b) + \frac{b}{r+b}V(r, b-1)\right\}$$

Starting with $V(r, 0) = r$ and $V(0, b) = 0$, the optimality equation can be utilized to obtain the desired value $V(n, m)$.   □

In some problems a reward is only earned when the problem ends.

Example 12.1c Suppose you can make up to n bets in sequence. At each bet you choose a stake amount s, which can be any nonnegative value less than or equal to your current fortune, and the result of the bet is that the amount sY is returned to you, where Y is a nonnegative random variable with a known distribution. Your objective is to maximize the expected value of the logarithm of your final fortune after n bets have taken place. Determine the optimal policy.

Solution. To begin, note that the state is your current fortune. So, let $V_n(x)$ be the maximal expected logarithm of your final fortune if your current fortune is x and n bets remain. Also, let the decision be the fraction of your wealth to stake. Because your fortune after betting the amount αx is $\alpha x Y + x - \alpha x = x(\alpha Y + 1 - \alpha)$, and $n - 1$ bets remain, the optimality equation becomes

$$V_n(x) = \max_{0 \le \alpha \le 1} E[V_{n-1}(x(\alpha Y + 1 - \alpha))]$$

Because $V_0(x) = \log(x)$, the preceding gives that

$$
\begin{aligned}
V_1(x) &= \max_{0 \le \alpha \le 1} E[\log(x(\alpha Y + 1 - \alpha))] \\
&= \log(x) + \max_{0 \le \alpha \le 1} E[\log(\alpha Y + 1 - \alpha)] \\
&= \log(x) + C
\end{aligned}
$$

where

$$C = \max_{0 \le \alpha \le 1} E[\log(\alpha Y + 1 - \alpha)]$$

Moreover, if we let

$$\alpha^* = \arg\max_{\alpha} E[\log(\alpha Y + 1 - \alpha)]$$

be the value of α that maximizes $E[\log(\alpha Y + 1 - \alpha)]$, then the optimal policy when only one bet can be made is to bet $\alpha^* x$ if your current wealth is x.

Now suppose your current fortune is x and two bets remain. Then the maximal expected logarithm of your final fortune is

$$
\begin{aligned}
V_2(x) &= \max_{0 \leq \alpha \leq 1} E[V_1(x(\alpha Y + 1 - \alpha))] \\
&= \max_{0 \leq \alpha \leq 1} E[\log(x(\alpha Y + 1 - \alpha)) + C] \\
&= \log(x) + C + \max_{0 \leq \alpha \leq 1} E[\log(\alpha Y + 1 - \alpha)] \\
&= \log(x) + 2C
\end{aligned}
$$

and it is once again optimal to stake the fraction α^* of your total wealth. Indeed, it is easy to see by using mathematical induction that

$$
V_n(x) = \log(x) + nC
$$

and that it is optimal, no matter how many bets remain, to always stake the fraction α^* of your total wealth.   □

12.2   Infinite Time Models

One is often interested in stochastic dynamic programming problems in which one wants to maximize the total expected reward earned over an infinite time horizon. That is, if the problem begins at time 0 and if X_n is the state at time n and A_n is the action chosen at time n, we are often interested in choosing the policy π that maximizes

$$
V_\pi(x) = E_\pi\left[\sum_{n=0}^{\infty} r(X_n, A_n) \Big| X_0 = x\right]
$$

where a policy π is a rule for choosing actions and we use the notation E_π to indicate that we are taking the expectation under the assumption that policy π is employed. Whereas the preceding, being the expected value of an infinite sum, may not be well defined or necessarily finite, we will suppose that the nature of the problem is such that it is well defined and finite. For instance, if we suppose that the one stage rewards $r(x, a)$ are bounded, say $|r(x, a)| < M$, and assume a discount factor β for which $0 \leq \beta < 1$, then the expected total discounted cost of a policy π would be bounded by $\frac{M}{1-\beta}$.

If we let

$$V(x) = \max_{\pi} V_{\pi}(x)$$

then $V(x)$ is the optimal value function, and satisfies the optimality equation

$$V(x) = \max_{a}\{r(x, a) + E[V(Y(x, a))]\}$$

Example 12.2a *An Optimal Asset Selling Problem* Suppose you receive an offer each day for an asset you desire to sell. When the offer is received, you must pay a cost $c > 0$ and then decide whether to accept or to reject the offer. Assuming that successive offers are independent with probability mass function $p_j = P(\text{offer is } j)$, $j \geq 0$, the problem is to determine the policy that maximizes the expected net return. Because the state is the current offer, let $V(i)$ denote the maximal additional net return from here on given that an offer of i has just been received. Now, if you accept the offer, then you receive the amount $-c + i$ and the problem ends. On the other hand, if you reject the offer, then you must pay c and wait for the next offer; if the next offer is j, then your maximal expected return from that point on would be $V(j)$. Because the next offer will equal j with probability p_j, it follows that the maximal expected net return if the offer of i is rejected is $-c + \sum_j p_j V(j)$. Because the maximum expected net return is the maximum of the maximum in the two cases, we see that the optimality equation is

$$V(i) = \max\left\{-c + i, \quad -c + \sum_j p_j V(j)\right\}$$

or, with $v = \sum_j p_j V(j)$,

$$V(i) = -c + \max\{i, \quad v\}$$

It follows from the preceding that the optimal policy is to accept offer i if and only if it is at least v. To determine v, note that

$$V(i) = \begin{cases} -c + v, & \text{if } i \leq v \\ -c + i, & \text{if } i > v \end{cases}$$

Hence,

$$v = \sum_i p_i V(i)$$

$$= -c + \sum_{i \le v} v p_i + \sum_{i > v} i p_i$$

Therefore, using that $\sum_i p_i = 1$, the preceding yields that

$$v \sum_{i > v} p_i = -c + \sum_{i > v} i p_i$$

or

$$\sum_{i > v} (i - v) p_i = c$$

or

$$c = \sum_i (i - v)^+ p_i$$

Hence, with X being a random variable having the distribution of an offer, the preceding states that

$$c = E[(X - v)^+] \tag{12.2}$$

That is, v is that value that makes $E[(X - v)^+]$ equal to c. (In most cases, v will have to be numerically determined.) The optimal policy is to accept the first offer that is at least v. Also, because $v = \sum_i p_i V(i)$ it follows that v is the maximum expected net return before the initial offer is received. $\square$

In most cases – such as when we have bounded rewards and a discount factor – the optimal value function V will be the limit of the n stage optimal value functions. That is, we would have that

$$V(x) = \lim_{n \to \infty} V_n(x)$$

This relationship can often be used to prove properties of the optimal value function by first using mathematical induction to prove that those properties are true for the optimal n stage returns and then letting n go to infinity. This is illustrated by our next example.

Example 12.2b *A Machine Replacement Model* Suppose that at the beginning of each period a machine is evaluated to be in some state $i, i = 0, \ldots, M$. After the evaluation, one must decide whether to pay the amount R and replace the machine or leave it alone. If the machine is replaced, then a new machine, whose state is 0, will be in place at the beginning of the next period. If a machine in state i is not replaced, then at the beginning of the next time period that machine will be in state j with probability $P_{i,j}$. Suppose that an operating cost $C(i)$ is incurred whenever the machine in use is evaluated as being in state i. Assume a discount factor $0 < \beta < 1$ and that our objective is to minimize the total expected discounted cost over an infinite time horizon.

If we let $V(i)$ denote the minimal expected discounted cost given that we start in state i, then the optimality equation is

$$V(i) = C(i) + \min \left\{ R + \beta V(0), \quad \beta \sum_j P_{i,j} V(j) \right\}$$

which follows because if we replace, then we incur an immediate cost of $C(i) + R$, and as the next state would be state 0, the minimal expected additional cost from then on would be $\beta V(0)$. On the other hand, if we do not replace, then our immediate cost is $C(i)$, and the best we could do if the next state were j would be $\beta V(j)$, showing that the minimal expected total discounted costs if we continue in state i is $C(i) + \beta \sum_j P_{i,j} V(j)$. Moreover, the policy that replaces a machine in state i if and only if

$$\beta \sum_j P_{i,j} V(j) \geq R + \beta V(0)$$

is an optimal policy.

Suppose we wanted to determine conditions that imply that $V(i)$ is increasing in i. One condition we might want to assume is that the operating costs $C(i)$ are increasing in i. So, let us make

Assumption 1 $C(i + 1) \geq C(i), \quad i \geq 0$.

However, after some thought it is easy to see that Assumption 1 by itself would not imply that $V(i)$ increases in i. For instance, even if we assume that $C(10) < C(11)$, it might be that state 11 is preferable to state 10, because even though it has a higher operating cost than state 10, it may be more likely to get you to a better state. So to rule this out, we shall suppose that $N(i)$, the next state of a not replaced machine that is currently in state i, is stochastically increasing in i. That is, we will make

Assumption 2 $N_{i+1} \geq_{st} N_i$, $i \geq 0$.

where $N_{i+1} \geq_{st} N_i$, means that $P(N_{i+1} \geq k) \geq P(N_i \geq k)$ for all k, which can be written as $\sum_{j \geq k} P_{i+1,j} \geq \sum_{j \geq k} P_{i,j}$ for all k. Moreover, by Proposition 10.1.1 of Section 10.1, Assumption 2 is equivalent to

Assumption 2 $E[h(N_i)]$ increases in i whenever h is an increasing function.

We now prove the following.

Theorem 12.1.1 *Under Assumptions* 1 *and* 2,

(a) $V(i)$ *is increasing in* i.
(b) *For some* $0 \leq i^* \leq \infty$, *the policy that replaces when in state i if and only if $i \geq i^*$ is an optimal policy.*

Proof. Let $V_n(i)$ denote the minimal expected discounted costs over an n-period problem that starts with a machine in state i. Then

$$V_n(i) = C(i) + \min \left\{ R + \beta V_{n-1}(0), \quad \beta \sum_j P_{i,j} V_{n-1}(j) \right\}, \quad n \geq 1$$

(12.3)

We now argue, using mathematical induction, that $V_n(i)$ is increasing in i for all n. Because $V_1(i) = C(i)$, it follows from Assumption 1 that the result is true when $n = 1$. So assume that $V_{n-1}(i)$ is increasing in i, and note that by Assumption 2 this implies that $E[V_{n-1}(N_i)]$ increases in i. But $E[V_{n-1}(N_i)] = \sum_j P_{i,j} V_{n-1}(j)$. Thus, from (12.3), it follows on using Assumption 1 that $V_n(i)$ increases in i, which completes the induction proof. Because $V(i) = \lim_{n \to \infty} V_n(i)$, we see that $V(i)$ increases in i.

We prove (b) by using that the optimal policy is to replace in state i if and only if

$$\beta \sum_j P_{i,j} V_{(j)} \geq R + \beta V(0)$$

which can be written as

$$E[V(N_i)] \geq \frac{R + \beta V(0)}{\beta}$$

But $E[V(N_i)]$ is, by part (a) and Assumption 2, an increasing function of i. Hence, it we let

$$i^* = \min \left\{ i : E[V(N_i)] \geq \frac{R + \beta V(0)}{\beta} \right\}$$

it follows that $E[V(N_i)] \geq \frac{R+\beta V(0)}{\beta}$ if and only if $i \geq i^*$. $\qquad\square$

12.3   Optimal Stopping Problems

An optimal stopping problem is a two-action problem. When in state x, one can either pay $c(x)$ and continue to the next state $Y(x)$, whose distribution depends only on x, or one can elect to stop, in which case one earns a final reward $r(x)$ and the problem ends. Letting $V(x)$ denote the maximal expected net additional return given that the current state is x, the optimality equation is

$$V(x) = \max\{r(x), \ -c(x) + E[V(Y(x))]\}$$

If the state space is the set of integers, then, with $P_{i,j}$ denoting the probability of going from state i to state j if one decides not to stop in state i, we can rewrite the preceding as

$$V(i) = \max \left\{ r(i), \ -c(i) + \sum_j P_{i,j} V(j) \right\}$$

Let $V_n(i)$ denote the maximal expected net return given that the current state is i and given that one is only allowed to go at most n additional time periods before stopping. Then, by the usual argument

$$V_0(i) = r(i)$$

and

$$V_n(i) = \max \left\{ r(i), \quad -c(i) + \sum_j P_{i,j} V_{n-1}(j) \right\}$$

Because having additional time periods before one must stop cannot hurt, it follows that $V_n(i)$ increases in n, and also that $V_n(i) \leq V(i)$.

Definition   If $\lim_{n \to \infty} V_n(i) = V(i)$, the stopping problem is said to be *stable*.

Most, though not all, stopping-rule problems that arise are stable. A sufficient condition for the stopping problem to be stable is the existence of constants $c > 0$ and $r < \infty$ such that

$$c(x) > c \quad \text{and} \quad r(x) < r \quad \text{for all } x$$

A policy that often has good results in optimal stopping problems is the *one-stage lookahead policy*, a policy that calls for stopping in state i if stopping would give a return that is at least as large as the expected return that would be obtained by continuing for exactly one more period and then stopping. That is, if we let

$$B = \left\{ i : r(i) \geq -c(i) + \sum_j P_{i,j} r(j) \right\}$$

be the set of states for which immediate stopping (which results in a final return $r(i)$) is at least as good as going exactly one more period and then stopping (which results in an expected additional return of $-c(i) + \sum_j P_{i,j} r(j)$), then the one-stage lookahead policy is the policy that stops when the current state i is in B and continues when it is not in B.

We now show for stable optimal stopping problems that if the set of states B is closed, in the sense that if the current state is in B and one chooses to continue then the next state will necessarily also be in B, then the one state lookahead policy is an optimal policy.

Theorem 12.3.1   *If the problem is stable and $P_{i,j} = 0$ for $i \in B$, $j \notin B$, then the one stage lookahead policy is an optimal policy.*

Proof. Note first that it cannot be optimal to stop in state i when $i \notin B$. This is so because better than stopping is to continue exactly one additional stage and then stop. So we need to prove that it is optimal to stop in state i when $i \in B$. That is, we must show that

$$V(i) = r(i), \quad i \in B \tag{12.4}$$

We prove this by showing, by mathematical induction, that for all n

$$V_n(i) = r(i), \quad i \in B$$

Because $V_0(i) = r(i)$, the preceding is true when $n = 0$. So assume that $V_{n-1}(i) = r(i)$ for all $i \in B$. Then, for $i \in B$

$$V_n(i) = \max \left\{ r(i), \ -c(i) + \sum_j P_{i,j} V_{n-1}(j) \right\}$$

$$= \max \left\{ r(i), \ -c(i) + \sum_{j \in B} P_{i,j} V_{n-1}(j) \right\} \quad \text{(since } B \text{ is closed)}$$

$$= \max \left\{ r(i), \ -c(i) + \sum_{j \in B} P_{i,j} r(j) \right\} \quad \begin{array}{l} \text{(by the induction} \\ \text{assumption)} \end{array}$$

$$= r(i)$$

where the final equality followed because $i \in B$. Hence, $V_n(i) = r(i)$ for $i \in B$, which yields (12.4) by stability, completing the proof. □

Example 12.3a Consider a burglar each of whose attempted burglaries is successful with probability p. If successful, the amount of loot earned is j with probability p_j, $j = 0, \dots, m$. If unsuccessful, the burglar is caught and loses everything he has accumulated to that time, and the problem ends. The burglar's problem is to decide whether to attempt another burglary or to stop and enjoy his accumulated loot. Find the optimal policy.

Solution. The state is the total loot so far collected. Now, if the current total loot is i and the burglar decides to stop, then he receives a reward i and the problem ends; on the other hand, if he decides to continue,

· then if successful the new state will be $i + j$ with probability p_j. Hence, if $V(i)$ is the burglar's maximal expected reward given that the current state is i, then the optimality equation is

$$V(i) = \max \left\{ i, \ p \sum_j p_j V(i + j) \right\}$$

The one-stage lookahead policy calls for stopping in state i if $i \in B$ where

$$B = \left\{ i : i \geq p \sum_j p_j(i + j) \right\}$$

That is, with $\mu = \sum_j jp_j$ denoting the expected return from a successful burglary,

$$B = \{i : i \geq p(i + \mu)\} = \left\{ i : i \geq \frac{p\mu}{1 - p} \right\}$$

Because the state cannot decrease (unless the burglar is caught and then no additional decisions are needed), it follows that B is closed, and so the one-stage lookahead policy that stops when the total loot is at least $\frac{p\mu}{1-p}$ is an optimal policy.   □

One-stage lookahead results give us an intuitive way of understanding the asset selling result of Example 12.2a.

Example 12.3b Letting $E[X]$ be the expected value of a new offer, the one-stage lookahead policy of Example 12.2a calls for accepting an offer j if $j \in B$, where

$$B = \{j : j \geq -c + E[X]\}$$

Because B is not a closed set of states (because successive offers need not be increasing), the one-stage lookahead policy would not necessarily be an optimal policy. However, suppose we change the problem by allowing the seller to be able to recall any past offer. That is suppose that a rejected offer is not lost, but may be accepted at any future time. In this case, the state after a new offer is observed would be the maximum offer ever received. Now, if j is the current state, then the selling price if we go exactly one more stage is $j + (X - j)^+$ where X is the offer in the final stage. Hence, the set of stopping states of the one-stage lookahead policy is

$$B = \{j : j \geq j + E[(X - j)^+] - c\} = \{j : E[(X - j)^+] \leq c\}$$

Because $E[(X - j)^+]$ is a decreasing function of j and because the state, being the maximum offer so far received, cannot decrease, it follows that B is a closed set of states. Hence, the one-stage lookahead policy is optimal in the recall problem. Now, if we let v be such that $E[(X - v)^+] = c$, then the one-stage lookahead policy in the recall problem is to accept the first offer that is at least v. However, because this policy can be employed even when no recall of past offers is allowed, it follows that it is also an optimal policy when no recall of past offers is allowed. (If it were not an optimal policy for the no-recall problem, then it would follow that the maximum expected net return in the no-recall problem would be strictly larger than in the recall problem, which clearly is not possible.) □

Our next example yields an interesting and surprising result about the mean number of times two players compete against each other in a multiple-player tournament in which each game involves two players.

Example 12.3b Consider a tournament involving k players, in which player i, $i = 1, \ldots, k$, starts with an initial fortune of $n_i > 0$. In each period, two of the players are chosen to play a game. The game is equally likely to be won by either player, and the winner of the game receives 1 from the loser. A player whose fortune drops to 0 is eliminated, and the tournament continues until one player has the entire fortune of $\sum_{i=1}^{k} n_i$. For specified players i and j we are interested in $E[N_{i,j}]$, where $N_{i,j}$ is the number of games in which i plays j.

To determine the mean number of times that i plays j, we set up a stopping-rule problem as follows. Suppose that immediately after the two players have been chosen for a game (and note that we have not yet specified how the players are chosen), we can either stop and receive a final reward equal to the product of the current fortunes of players i and j, or we can continue. If we continue, then we receive a reward of 1 in that period if the two contestants are i and j, or a reward of 0 if the contestants are not i and j. Suppose the current fortunes of i and j are n

and m. Then stopping at this time will yield a final reward of nm. On the other hand, if we continue for one additional period and then stop, we will receive a total reward of nm if i and j are not the competitors in the current round (because we receive 0 during that period and then nm when we stop the following period), and we will receive the expected amount

$$1 + \frac{1}{2}(n + 1)(m - 1) + \frac{1}{2}(n - 1)(m + 1) = nm$$

if i and j are the competitors. Hence, in all cases the return from immediately stopping is exactly the same as the expected return from going exactly one more period and then stopping. Thus, the one-stage lookahead policy always calls for stopping, and as its set of stopping states is thus closed, it follows that it is an optimal policy. But because continuing on for an additional period and then stopping yields the same expected return as immediately stopping, it follows that always continuing is also optimal. But the total return from the policy that always continues is $N_{i,j}$, the number of times that i and j play each other. Because $n_i n_j$ is the return from immediately stopping, we see that $E[N_{i,j}] = n_i n_j$. Moreover, interestingly enough, this result is true no matter how the contestants in each round are chosen. $\qquad \square$

12.4   Exercises

Exercise 12.1 To be successful, you need to build a specified number of working machines, and you have a specified number of dollars to accomplish the task. You must spend an integral amount on each machine, and if you spend j, then the machine will work with probability $p(j)$, $j = 0, 1, \ldots$, where $p(0) = 0$. The machines are to be built sequentially, and when a machine is completed, you immediately learn whether or not it works. Let $V_k(n)$ be the maximal probability of being successful given that you have n to spend and still need k working machines.

(a) Derive an equation for $V_k(n)$.
(b) Find the optimal policy and maximal probability of being able to build two working machines when you have 4 dollars, and $p(1) = 0.2$, $p(2) = 0.4$, $p(3) = 0.6$, $p(4) = 1$.

Exercise 12.2 In Example 12.1b, find the optimal strategy and the optimal value when the urn contains three red and four blue balls.

Exercise 12.3 Complete the proof in Example 12.1c that $V_n(x) = \log(x) + nC$ and that the optimal policy is to always bet the fraction α^* of your total wealth. Also, show that $\alpha^* = 0$ if $E[Y] \leq 1$.

Exercise 12.4 Find the optimal policy in Example 12.4 when there is discount factor β.

Exercise 12.5 Each time you play a game you either win or lose. Before playing each game, you must decide how much to invest in that game, with the amount determining your probability of winning. Specifically, if you invest x, then you will win that game with probability $p(x)$, where $p(x)$ is an increasing function of x. Suppose you must invest at least 1 in each game, and that you must continue to play until you have won n games in a row.

Let V_k denote the minimal expected cost incurred until you have won k games in a row.

(a) Explain the equation

$$V_k = \min_{x \geq 1}\{V_{k-1} + x + (1 - p(x))V_k\}$$

(b) Show that $V_k, k \geq 1$ are recursively determined by

$$V_1 = \min_{x \geq 1} \frac{x}{p(x)}$$

$$V_k = \min_{x \geq 1} \frac{V_{k-1} + x}{p(x)}, \quad k = 2, \ldots, n$$

(c) In terms of the values $V_k, k \geq 1$, what is the optimal policy?

Exercise 12.6 At each stage, one can either pay 1 and receive a coupon that is equally likely to be any of n types, or one can stop and receive a final reward of jr if one's current collection of coupons contains exactly j distinct types. Thus, for instance, if one stops after having previously obtained six coupons whose successive types were 2, 4, 2, 5, 4, 3, then one would have earned a net return of $4r - 6$. The objective is to maximize the expected net return.

We want to solve this as a dynamic programming problem.

(a) What are the states and actions?
(b) Define the optimal value function and give the optimality equation.
(c) Give the one-stage lookahead policy.
(d) Is the one-stage lookahead policy an optimal policy? Explain.
    Now suppose that each coupon obtained is type i with probability p_i, $\sum_{i=1}^{n} p_i = 1$.
(e) Give the states in this case.
(f) Give the one-stage lookahead policy and explain whether it is an optimal policy.

Exercise 12.7   In Example 12.1b, is the one-stage lookahead policy an optimal policy? If not optimal, do you think it would be a good policy?

REFERENCE

[1] Ross, S. M. (1983). *Introduction to Stochastic Dynamic Programming*, Academic Press.

13. Exotic Options

13.1 Introduction

The options we have so far considered are sometimes called "vanilla" options to distinguish them from the more exotic options, whose prevalence has increased in recent years. Generally speaking, the value of these options at the exercise time depends not only on the security's price at that time but also on the price path leading to it. In this chapter we introduce three of these exotic-type options – barrier options, Asian options, and lookback options – and show how to use Monte Carlo simulation methods efficently to determine their geometric Brownian motion risk-neutral valuations. In the final section of this chapter we present an explicit formula for the risk-neutral valuation of a "power" call option, whose payoff when exercised is the amount by which a specified power of the security's price at that time exceeds the exercise price.

13.2 Barrier Options

To define a European barrier call option with strike price K and exercise time t, a barrier value v is specified; depending on the type of barrier option, the option either becomes alive or is killed when this barrier is crossed. A *down-and-in* barrier option becomes alive only if the security's price goes below v before time t, whereas a *down-and-out* barrier option is killed if the security's price goes below v before time t. In both cases, v is a specified value that is less than the initial price s of the security. In addition, in most applications, the barrier is considered to be breached only if an end-of-day price is lower than v; that is, a price below v that occurs in the middle of a trading day is not considered to breach the barrier. Now, if one owns both a down-and-in and a down-and-out call option, both with the same values of K and t, then exactly one option will be in play at time t (the down-and-in option if the barrier is breached and the down-and-out otherwise); hence, owning both is equivalent to owning a vanilla option with exercise time t and exercise price K. As a result, if $D_i(s, t, K)$ and $D_o(s, t, K)$ represent, respectively, the risk-neutral present values of owning the down-and-in and the down-and-out call options, then

$$D_i(s, t, K) + D_o(s, t, K) = C(s, t, K),$$

where $C(s, t, \dot{K})$ is the Black–Scholes valuation of the call option given by Equation (7.2). As a result, determining either one of the values $D_i(s, t, K)$ or $D_o(s, t, K)$ automatically yields the other.

There are also *up-and-in* and *up-and-out* barrier call options. The up-and-in option becomes alive only if the security's price exceeds a barrier value v, whereas the up-and-out is killed when that event occurs. For these options, the barrier value v is greater than the exercise price K. Since owning both these options (with the same t and K) is equivalent to owning a vanilla option, we have

$$U_i(s, t, K) + U_o(s, t, K) = C(s, t, K),$$

where U_i and U_o are the geometric Brownian motion risk-neutral valuations of (resp.) the up-and-in and the up-and-out call options, and C is again the Black–Scholes valuation.

13.3   Asian and Lookback Options

Asian options are options whose value at the time t of exercise is dependent on the *average price* of the security over at least part of the time between 0 (when the option was purchased) and the time of exercise. As these averages are usually in terms of the end-of-day prices, let N denote the number of trading days in a year (usually taken equal to 252), and let

$$S_d(i) = S(i/N)$$

denote the security's price at the end of day i. The most common Asian-type call option is one in which the exercise time is the end of n trading days, the strike price is K, and the payoff at the exercise time is

$$\left(\sum_{i=1}^{n} \frac{S_d(i)}{n} - K \right)^+.$$

Another Asian option variation is to let the average price be the strike price; the final value of this call option is thus

$$\left(S_d(n) - \sum_{i=1}^{n} \frac{S_d(i)}{n} \right)^+$$

when the exercise time is at the end of trading day n.

Another type of exotic option is the *lookback option,* whose strike price is the minimum end-of-day price up to the option's exercise time. That is, if the exercise time is at the end of n trading days, then the payoff at exercise time is

$$S_d(n) - \min_{i=1,\ldots,n} S_d(i).$$

Another lookback option variation is to substitute the maximum end-of-day price for the final price in the payoff of a call option with strike K. That is, the payoff at exercise time would be

$$(\max_{i=1,\ldots,n} S_d(i) - K)^+$$

Because their final payoffs depend on the end-of-day price path followed, there are no known exact formulas for the risk-neutral valuations of barrier, Asian, or lookback options. However, fast and accurate approximations are obtainable from efficient Monte Carlo simulation methods.

13.4   Monte Carlo Simulation

Suppose we want to estimate θ, the expected value of some random variable Y:

$$\theta = E[Y].$$

Suppose, in addition, that we are able to genererate the values of independent random variables having the same probability distribution as does Y. Each time we generate a new value, we say that a simulation "run" is completed. Suppose we perform k simulation runs and so generate the values of (say) $Y_1, Y_2, \ldots, Y_k$. If we let

$$\bar{Y} = \frac{1}{k} \sum_{i=1}^{k} Y_i$$

be their arithmetic average, then $\bar{Y}$ can be used as an estimator of θ. Its expected value and variance are as follows. For the expected value we have

$$E[\bar{Y}] = \frac{1}{k} \sum_{i=1}^{k} E[Y_i] = \theta.$$

Also, letting

$$v^2 = \text{Var}(Y),$$

we have that

$$\text{Var}(\bar{Y}) = \text{Var}\left(\frac{1}{k} \sum_{i=1}^{k} Y_i\right)$$

$$= \frac{1}{k^2} \text{Var}\left(\sum_{i=1}^{k} Y_i\right)$$

$$= \frac{1}{k^2} \sum_{i=1}^{k} \text{Var}(Y_i) \quad \text{(by independence)}$$

$$= v^2/k.$$

Also, it follows from the central limit theorem that, for large k, $\bar{X}$ will have an approximately normal distribution. Hence, as a normal random variable tends not to be too many standard deviations (equal to the square root of its variance) away from its mean, it follows that if $v/\sqrt{k}$ is small then $\bar{X}$ will tend to be near θ. (For instance, since more than 95% of the time a normal random variable is within two standard deviations of its mean, we can be 95% certain that the generated value of $\bar{X}$ will be within $2v/\sqrt{k}$ of θ.) Hence, when k is large, $\bar{X}$ will tend to be a good estimator of θ. (To know exactly how good, we would use the generated sample variance to estimate v^2.) This approach to estimating an expected value is known as *Monte Carlo simulation*.

13.5   Pricing Exotic Options by Simulation

Suppose that the nominal interest rate is r and that the price of a security follows the risk-neutral geometric Brownian motion; that is, it follows a geometric Brownian motion with variance parameter σ^2 and drift parameter μ, where

$$\mu = r - \sigma^2/2.$$

Let $S_d(i)$ denote the price of the security at the end of day i, and let

$$X(i) = \log\left(\frac{S_d(i)}{S_d(i-1)}\right).$$

Successive daily price ratio changes are independent under geometric Brownian motion, so it follows that $X(1), \ldots, X(n)$ are independent normal random variables, each having mean μ/N and variance σ^2/N (as before, N denotes the number of trading days in a year). Therefore, by generating the values of n independent normal random variables having this mean and variance, we can construct a sequence of n end-of-day prices that have the same probabilities as ones that evolved from the risk-neutral geometric Brownian motion model. (Most computer languages and almost all spreadsheets have built-in utilities for generating the values of standard normal random variables; multiplying these by $\sigma/\sqrt{N}$ and then adding μ/N gives the desired normal random variables.)

Suppose we want to find the risk-neutral valuation of a down-and-in barrier option whose strike price is K, barrier value is v, initial value is $S(0) = s$, and exercise time is at the end of trading day n. We begin by generating n independent normal random variables with mean μ/N and variance σ^2/N. Set them equal to $X(1), \ldots, X(n)$, and then determine the sequence of end-of-day prices from the equations

$$S_d(0) = s,$$

$$S_d(1) = S_d(0)e^{X(1)},$$

$$S_d(2) = S_d(1)e^{X(2)};$$

$$\vdots$$

$$S_d(i) = S_d(i-1)e^{X(i)};$$

$$\vdots$$

$$S_d(n) = S_d(n-1)e^{X(n)}.$$

In terms of these prices, let I equal 1 if an end-of-day price is ever below the barrier v, and let it equal 0 otherwise; that is,

$$I = \begin{cases} 1 & \text{if } S_d(i) < v \text{ for some } i = 1, \ldots, n, \\ 0 & \text{if } S_d(i) \geq v \text{ for all } i = 1, \ldots, n. \end{cases}$$

Then, since the down-and-in call option will be alive only if $I = 1$, it follows that the time-0 value of its payoff at expiration time n is

payoff of the down-and-in call option $= e^{-rn/N} I (S_d(n) - K)^+$.

Call this payoff Y_1. Repeating this procedure an additional $k - 1$ times yields $Y_1, \ldots, Y_k$, a set of k payoff realizations. We can then use their average as an estimate of the risk-neutral geometric Brownian motion valuation of the barrier option.

Risk-neutral valuations of Asian and lookback call options are similarly obtained. As in the preceding, we first generate the values of $X(1), \ldots, X(n)$ and use them to compute $S_d(1), \ldots, S_d(n)$. For an Asian option, we then let

$$Y = e^{-rn/N} \left(\sum_{i=1}^{n} \frac{S_d(i)}{n} - K \right)^+$$

if the strike price is fixed at K and the payoff is based on the average end-of-day price, or we let

$$Y = e^{-rn/N} \left(S_d(n) - \sum_{i=1}^{n} \frac{S_d(i)}{n} \right)^+$$

if the average end-of-day price is the strike price. In the case of a lookback option, we would let

$$Y = e^{-rn/N} \left(S_d(n) - \min_i S_d(i) \right).$$

Repeating this procedure an additional $k - 1$ times and then taking the average of the k values of Y yields the Monte Carlo estimate of the risk-neutral valuation.

13.6   More Efficient Simulation Estimators

In this section we show how the simulation of valuations of Asian and lookback options can be made more efficient by the use of control and antithetic variables, and how the valuation simulations of barrier options can be improved by a combination of the variance reduction simulation techniques of conditional expectation and importance sampling.

13.6.1   *Control and Antithetic Variables in the Simulation of Asian and Lookback Option Valuations*

Consider the general setup where one plans to use simulation to estimate

$$\theta = E[Y].$$

Suppose that, in the course of generating the value of the random variable Y, we also learn the value of a random variable V whose mean value is known to be $\mu_V = E[V]$. Then, rather than using the value of Y as the estimator, we can use one of the form

$$Y + c(V - \mu_V),$$

where c is a constant to be specified. That this quantity also estimates θ follows by noting that

$$E[Y + c(V - \mu_V)] = E[Y] + cE[V - \mu_V] = \theta + c(\mu_V - \mu_V) = \theta.$$

The best estimator of this type is obtained by choosing c to be the value that makes $\mathrm{Var}(Y + c(V - \mu_V))$ as small as possible. Now,

$$
\begin{aligned}
\mathrm{Var}(Y + c(V - \mu_V)) &= \mathrm{Var}(Y + cV) \\
&= \mathrm{Var}(Y) + \mathrm{Var}(cV) + 2\,\mathrm{Cov}(Y, cV) \\
&= \mathrm{Var}(Y) + c^2\,\mathrm{Var}(V) + 2c\,\mathrm{Cov}(Y, V).
\end{aligned}
\tag{13.1}
$$

If we differentiate Equation (13.1) with respect to c, set the derivative equal to 0, and solve for c, then it follows that the value of c that minimizes $\mathrm{Var}(Y + c(V - \mu_V))$ is

$$c^* = -\frac{\mathrm{Cov}(Y, V)}{\mathrm{Var}(V)}.$$

Substituting this value back into Equation (13.1) yields

$$\mathrm{Var}(Y + c^*(V - \mu_V)) = \mathrm{Var}(Y) - \frac{\mathrm{Cov}^2(Y, V)}{\mathrm{Var}(V)}. \tag{13.2}$$

Dividing both sides of this equation by $\mathrm{Var}(Y)$ shows that

$$\frac{\mathrm{Var}(Y + c^*(V - \mu_V))}{\mathrm{Var}(Y)} = 1 - \mathrm{Corr}^2(Y, V),$$

where

$$\mathrm{Corr}(Y, V) = \frac{\mathrm{Cov}(Y, V)}{\sqrt{\mathrm{Var}(Y)\,\mathrm{Var}(V)}}$$

is the correlation between Y and V. Hence, the variance reduction obtained when using the *control variable* V is $100\,\mathrm{Corr}^2(Y, V)$ percent.

The quantities $\mathrm{Cov}(Y, V)$ and $\mathrm{Var}(V)$, which are needed to determine c^*, are not usually known and must be estimated from the simulated data. If k simulation runs produce the output Y_i and V_i ($i = 1, \ldots, k$) then, letting

$$\bar{Y} = \sum_{i=1}^{k} \frac{Y_i}{k} \quad \text{and} \quad \bar{V} = \sum_{i=1}^{k} \frac{V_i}{k}$$

be the sample means, $\mathrm{Cov}(Y, V)$ is estimated by

$$\frac{\sum_{i=1}^{k}(Y_i - \bar{Y})(V_i - \bar{V})}{k - 1}$$

and $\mathrm{Var}(V)$ is estimated by the sample variance

$$\frac{\sum_{i=1}^{k}(V_i - \bar{V})^2}{k - 1}.$$

Combining the preceding estimators gives the estimator of c^*, namely,

$$\widehat{c^*} = -\frac{\sum_{i=1}^{k}(Y_i - \bar{Y})(V_i - \bar{V})}{\sum_{i=1}^{k}(V_i - \bar{V})^2},$$

and produces the following controlled simulation estimator of θ:

$$\frac{1}{k} \sum_{i=1}^{k} (Y_i + \widehat{c^*}(V_i - \mu_V)).$$

Let us now see how control variables can be gainfully employed when simulating Asian option valuations. Suppose first that the present value of the final payoff is

$$Y = e^{-rn/N}\left(\sum_{i=1}^{n} \frac{S_d(i)}{n} - K \right)^{+}.$$

It is clear that Y is strongly positively correlated with

$$V = \sum_{i=0}^{n} S_d(i),$$

so one possibility is to use V as a control variable. Toward this end, we must first determine $E[V]$. Because

$$E[S_d(i)] = e^{ri/N}S(0)$$

for a risk-neutral valuation, we see that

$$E[V] = E\left[\sum_{i=0}^{n} S_d(i)\right]$$

$$= \sum_{i=0}^{n} E[S_d(i)]$$

$$= S(0) \sum_{i=0}^{n} (e^{r/N})^i$$

$$= S(0)\frac{1 - e^{r(n+1)/N}}{1 - e^{r/N}}.$$

Another choice of control variable that could be used is the payoff from a vanilla option with the same strike price and exercise time. That is, we could let

$$V = (S_d(n) - K)^+$$

be the control variable.

A different variance reduction technique that can be effectively employed in this case is to use *antithetic variables.* This method generates the data $X(1), \ldots, X(n)$ and uses them to compute Y. However, rather than generating a second set of data, it re-uses the same data with the following changes:

$$X(i) \implies \frac{2(r - \sigma^2/2)}{N} - X(i).$$

That is, it lets the new value of $X(i)$ be $2(r - \sigma^2/2)/N$ minus its old value, for each $i = 1, \ldots, n$. (The new value of $X(i)$ will be negatively correlated with the old value, but it will still be normal with the same mean and variance.) The value of Y based on these new values is then computed, and the estimate from that simulation run is the average of the two Y values obtained. It can be shown (see [5]) that re-using the data in this manner will result in a smaller variance than would be obtained by generating a new set of data.

Now let us consider an Asian call option for which the strike price is the average end-of-day price; that is, the present value of the final payoff is

$$Y = e^{-rn/N}\left(S_d(n) - \sum_{i=1}^{n} \frac{S_d(i)}{n}\right)^+.$$

Recall that a simulation run consists of (a) generating $X(1), \ldots, X(n)$ independent normal random variables with mean $(r - \sigma^2/2)/N$ and variance σ^2/N, and (b) setting

$$S_d(i) = S(0)e^{X(1)+\cdots+X(i)}, \quad i = 1, \ldots, n.$$

Since the value of Y will be large if the latter values of the the sequence $X(1), X(2), \ldots, X(n)$ are among the largest (and small if the reverse is true), one could try a control variable of the type

$$V = \sum_{i=1}^{n} w_i X(i),$$

where the weights w_i are increasing in i. However, we recommend that one use all of the variables $X(1), X(2), \ldots, X(n)$ as control variables. That is, from each run one should consider the estimator

$$Y + \sum_{i=1}^{n} c_i\left(X(i) - \frac{r - \sigma^2/2}{N}\right).$$

Because the control variables are independent, it is easy to verify (see Exercise 13.4) that the optimal values of the c_i are

$$c_i = -\frac{\mathrm{Cov}(X(i), Y)}{\mathrm{Var}(X(i))}, \quad i = 1, \ldots, n;$$

these quantities can be estimated from the output of the simulation runs. We suggest this same approach in the case of lookback options also: again, use all of the variables $X(1), X(2), \ldots, X(n)$ as control variables.

13.6.2 *Combining Conditional Expectation and Importance Sampling in the Simulation of Barrier Option Valuations*

In Section 13.5 we presented a simulation approach for determining the expected value of the risk-neutral payoff under geometric Brownian motion of a down-and-in barrier call option. The $X(i)$ were generated and used to calculate the successive end-of-day prices and the resulting payoff from the option. We can improve upon this approach by noting that, in order for this option to become alive, at least one of the end-of-day prices must fall below the barrier. Suppose that with the generated data this first occurs at the end of day j, with the price at the end of that day being $S_d(j) = x < v$. At this moment the barrier option becomes alive and its worth is exactly that of an ordinary vanilla call option, given that the price of the security is x when there is time $(n - j)/N$ that remains before the option expires. But this implies that the option's worth is now $C(x, (n - j)/N, K)$. Consequently, it seems that we could (a) end the simulation run once an end-of-day price falls below the barrier, and (b) use the resulting Black–Scholes valuation as the estimator from this run. As a matter of fact, we can do this; the resulting estimator, called the *conditional expectation estimator,* can be shown to have a smaller variance than the one derived in Section 13.5.

The conditional expectation estimator can be further improved by making use of the simulation idea of *importance sampling.* Since many of the simulation runs will never have an end-of-day price fall below the barrier, it would be nice if we could first simulate the data from a set of probabilities that makes it more likely for an end-of-day price to fall below the barrier and then add a factor to compensate for these different probabilities. This is exactly what importance sampling does. It generates the random variables $X(1), X(2), \ldots$ from a normal distribution with mean $(r - \sigma^2/2)/N - b$ and variance σ^2/N, and it determines the first time that a resulting end-of-day price falls below the barrier. If the price first falls below the barrier at time j with price x, then the estimator from that run is

$$C(x, (n - j)/N, K) \exp\left\{ \frac{jb^2 N}{2\sigma^2} + \frac{Nb}{\sigma^2} \sum_{i=1}^{j} X_i - \frac{jb}{\sigma^2}\left(r - \frac{\sigma^2}{2}\right) \right\}$$

(see [6] for details); if the price never falls below the barrier then the estimator from that run is 0. The average of these estimators over many runs is the overall estimator of the value of the option. Of course, in order to implement this procedure one needs an appropriate choice of b. Probably the best approach to choosing b is empirical; do some small simulations in cases of interest, and see which value of b leads to a small variance. In addition, the choice

$$b = \frac{r - \sigma^2/2}{N} - \frac{2\log\left(\frac{S(0)}{v}\right) + \log\left(\frac{K}{S(0)}\right)}{n}$$

was shown (in [1]) to work well for a less efficient variation of our method.

13.7   Options with Nonlinear Payoffs

The standard call option has a payoff that, provided the security's price at exercise time is in the money, is a linear function of that price. However, there are more general options whose payoff is of the form

$$\left(h(S(t)) - K\right)^+,$$

where h is an arbitrary specified function, t is the exercise time, and K is the strike price. Whereas a simulation or a numerical procedure based on a multiperiod binomial approximation to geometric Brownian motion is often needed to determine the geometric Brownian motion risk-neutral valuations of these options, an exact formula can be derived when h is of the form

$$h(x) = x^\alpha.$$

Options having nonlinear payoffs $(S^\alpha(t) - K)^+$ are called *power options,* and α is called the power parameter.

Let $C_\alpha(s, t, K, \sigma, r)$ be the risk-neutral valuation of a power call option with power parameter α that expires at time t with an exercise price K, when the interest rate is r, the underlying security initially has price s, and the security follows a geometric Brownian motion with volatility σ. As usual, let $C(s, t, K, \sigma, r) = C_1(s, t, K, \sigma, r)$ be the Black–Scholes valuation. Also, let X be a normal random variable with mean $(r - \sigma^2/2)t$ and variance $\sigma^2 t$. Because e^X has the same probability distribution as does $S(t)/s$, it follows that

$$e^{rt}C(s, t, K, \sigma, r) = E[(S(t) - K)^+] = E[(se^X - K)^+]. \quad (13.3)$$

In addition, since $(S(t)/s)^\alpha = S^\alpha(t)/s^\alpha$ has the same distribution as does $e^{\alpha X}$, it follows that

$$E[(S^\alpha(t) - K)^+] = E[(s^\alpha e^{\alpha X} - K)^+]. \quad (13.4)$$

But since αX is a normal random variable with mean $\alpha(r - \sigma^2/2)t$ and variance $\alpha^2\sigma^2 t$, it follows from Equation (13.3) that if we let r_α and σ_α be such that

$$r_\alpha - \sigma_\alpha^2/2 = \alpha(r - \sigma^2/2) \quad \text{and} \quad \sigma_\alpha^2 = \alpha^2\sigma^2$$

then

$$e^{r_\alpha t}C(s^\alpha, t, K, \sigma_\alpha, r_\alpha) = E[(s^\alpha e^{\alpha X} - K)^+].$$

Hence, from Equation (13.4) we obtain that

$$e^{-rt}E[(S^\alpha(t) - K)^+]$$

$$= e^{-rt}e^{r_\alpha t}C(s^\alpha, t, K, \alpha\sigma, r_\alpha)$$

$$= \exp\{(\alpha(r - \sigma^2/2) + \alpha^2\sigma^2/2 - r)t\}C(s^\alpha, t, K, \alpha\sigma, r_\alpha)$$

$$= \exp\{(\alpha - 1)(r + \alpha\sigma^2/2)t\}C(s^\alpha, t, K, \alpha\sigma, r_\alpha).$$

That is,

$$C_\alpha(s, t, K, \sigma, r) = \exp\{(\alpha - 1)(r + \alpha\sigma^2/2)t\}C(s^\alpha, t, K, \alpha\sigma, r_\alpha),$$

where

$$r_\alpha = \alpha(r - \sigma^2/2) + \alpha^2\sigma^2/2.$$

13.8 Pricing Approximations via Multiperiod Binomial Models

Multiperiod binomial models can also be used to determine efficiently the risk-neutral geometric Brownian motion prices of certain exotic options. For instance, consider the down-and-out barrier call option having initial price s, strike price K, exercise time $t = n/N$ (where N is the number of trading days in a year), and barrier value v ($v < s$). To begin, choose an integer j, let $m = nj$, and let $t_k = kt/m$ ($k = 0, 1, ..., m$). We will consider each day as consisting of j periods and willl approximate using an m-period binomial model that supposes

$$S(t_{k+1}) = \begin{cases} uS(t_k) & \text{with probability } p, \\ dS(t_k) & \text{with probability } 1 - p, \end{cases}$$

where

$$u = e^{\sigma\sqrt{t/m}}, \quad d = e^{-\sigma\sqrt{t/m}},$$

$$p = \frac{1 + rt/m - d}{u - d}.$$

If i of the first k price movements are increases and $k - i$ are decreases, then the price at time t_k is

$$S(t_k) = u^i d^{k-i} s.$$

Letting $V_k(i)$ denote the expected payoff from the barrier call option given that the option is still alive at time t_k and that the price at time t_k is $S(t_k) = u^i d^{k-i} s$, we can approximate the expected present value payoff of the European barrier call option by $e^{-rt} V_0(0)$. The value of $V_0(0)$ can be obtained by working backwards. That is, we start with the identity

$$V_m(i) = (u^i d^{m-i} s - K)^+, \quad i = 0, ..., m,$$

to determine the values of $V_m(i)$ and then repeatedly use the following equation (initially with $k = m - 1$, and then decreasing its value by 1 after each interation):

$$V_k(i) = pV_{k+1}(i + 1) + (1 - p)W_{k+1}(i), \tag{13.5}$$

where

$$W_{k+1}(i) = \begin{cases} 0 & \text{if } u^i d^{k+1-i} s < v \text{ and } j \text{ divides } k + 1, \\ V_{k+1}(i) & \text{otherwise.} \end{cases}$$

Note that $W_{k+1}(i)$ is defined in this fashion because if j divides $k + 1$ then the period-$(k + 1)$ price is an end-of-day price and will thus kill the option if it is less than the barrier value.

If we wanted the risk-neutral price of a down-and-in call option then we could use an analogous procedure. Alternatively, we could use the preceding to determine the price of a down-and-out call option with the same parameters and then use the identity

$$D_i(s, t, K) + D_o(s, t, K) = C(s, t, K),$$

where D_i, D_o, and C refer to the risk-neutral price of (respectively) a down-and-in call option, a down-and-out call option, and a vanilla Black–Scholes call option.

Risk-neutral prices of other exotic options can also be approximated by multiperiod binomial models. However, the computational burden can be demanding. For instance, consider an Asian option whose strike price is the average of the end-of-day prices. To recursively determine the expected value of the final payoff given all that has occurred up to time t_k, we need to specify not only the price at time t_k but also the sum of the end-of-day prices up to that time. That is, in order to approximate an n-day call option with an n-period binomial model, we would need to recursively compute the values $V_k(i, x)$ equal to the expected final payoff given that the price after k periods is $u^i d^{k-i} s$ and that the sum of the first k prices is x. Since there can be as many as $\binom{k}{i}$ possible sums of the first k prices when i of them are increases, it can require a great deal of computation to obtain a good approximation. Generally speaking, we recommend the use of simulation to estimate the risk-neutral prices of most path-dependent exotic options.

13.9 Continuous Time Approximations of Barrier and Lookback Options

The no-arbitrage cost of barrier options, say an up and out barrier option, can also be approximated by considering a continuous time variation that declares the option dead if any (not just an end-of-day) price up to expiration time t exceeds the barrier value v. That is, the payoff at time t is $I(S(t) - K)^+$, where

$$I = \begin{cases} 1, & \text{if } \max_{0 \le w \le t} S(w) \le v \\ 0, & \text{if } \max_{0 \le w \le t} S(w) > v \end{cases}$$

To compute the expected present-value payoff under the risk-neutral geometric Brownian motion, we use its representation

$$S(w) = se^{X(w)}, \quad w \ge 0$$

where $s = S(0)$ and where $X(w)$, $w \geq 0$ is Brownian motion with drift parameter $\mu_r \equiv r - \sigma^2/2$ and variance parameter σ^2 that has $X(0) = 0$. Hence, letting $f_{X(t)}$ be the density of $X(t)$, a normal random variable with mean $\mu_r t$ and variance $t\sigma^2$, we obtain upon conditioning on $X(t)$ that

$$E[I(S(t) - K)^+] = E[I(se^{X(t)} - K)^+]$$

$$= \int_{-\infty}^{\infty} E[I(se^{X(t)} - K)^+ | X(t) = x] f_{X(t)}(x)\, dx$$

Letting $M(t) = \max_{0 \leq w \leq t} X(w)$, it follows that $I = 1$ if $se^{M(t)} \leq v$ and is equal to 0 otherwise. Using this and that the payoff of the option is necessarily 0 if $S(t) = se^{X(t)}$ is not between K and v, we see from the preceding that

$$E[I(S(t) - K)^+] = \int_{\ln(K/s)}^{\ln(v/s)} (se^x - K) E[I | X(t) = x] f_{X(t)}(x)\, dx$$

$$= \int_{\ln(K/s)}^{\ln(v/s)} (se^x - K) P(M(t) \leq \ln(v/s) | X(t) = x)$$

$$\times \frac{1}{\sqrt{2\pi t}\, \sigma} e^{-(x - t\mu_r)^2/2t\sigma^2}\, dx$$

Using Theorem 3.4.1 of Chapter 3, which gives the conditional distribution of $M(t)$ given the value of $X(t)$, the preceding integral can be explicitly determined. We leave the details to the interested reader.

Similar analysis to the preceding can be used to obtain explicit expressions for the expected present value returns from lookback options that use payoffs of the form $S(t) - \min_{0 \leq w \leq t} S(w)$ or $(\max_{0 \leq w \leq t} S(w) - K)^+$. The computation in the former case would first condition on $X(t)$ and would then use the conditional distribution of $\min_{0 \leq w \leq t} X(w)$ given $X(t)$. The computation in the latter case would just use the distribution of the maximum up to time t of a Brownian motion process.

13.10   Exercises

Exercise 13.1 Consider an American call option that can be exercised at any time up to time t; however, if it is exercised at time y (where $0 \leq y \leq t$) then the strike price is Ke^{uy} for some specified value of u. That is, the payoff if the call is exercised at time y $(0 \le y \le t)$ is

$$(S(y) - e^{uy}K)^+.$$

Argue that if $u \le r$ then the call should never be exercised early, where r is the interest rate.

Exercise 13.2 A lookback put option that expires after n trading days has a payoff equal to the maximum end-of-day price achieved by time n minus the price at time n. That is, the payoff is

$$\max_{0 \le i \le n} S_d(i) - S_d(n).$$

Explain how Monte Carlo simulation can be used efficiently to find the geometric Brownian motion risk-neutral price of such an option.

Exercise 13.3 In Section 13.6.1, it is noted that $V = (S_d(n) - K)^+$ can be used as a control variate. However, doing so requires that we know its mean; what is $E[V]$?

Exercise 13.4 Let $X_1, \ldots, X_n$ be independent random variables with expected values $E[X_i] = \mu_i$, and consider the following simulation estimator of $E[Y]$:

$$W = Y + \sum_{i=1}^{n} c_i(X_i - \mu_i).$$

(a) Show that

$$\text{Var}(W) = \text{Var}(Y) + \sum_{i=1}^{n} c_i^2 \, \text{Var}(X_i) + 2 \sum_{i=1}^{n} c_i \, \text{Cov}(Y, X_i).$$

(b) Use calculus to show that the values of $c_1, \ldots, c_n$ that minimize $\text{Var}(W)$ are

$$c_i = -\frac{\text{Cov}(Y, X_i)}{\text{Var}(X_i)}, \quad i = 1, \ldots, n.$$

Exercise 13.5 Perform a Monte Carlo simulation to estimate the risk-neutral valuation of some exotic option. Do it first without any attempts at variance reduction and then a second time with some variance reduction procedure.

Exercise 13.6 Give the equations that are needed when using a multi-period binomial model to approximate the risk-neutral price of a down-and-in barrier call option.

Exercise 13.7 Explain how you can approximate the risk-neutral price of a down-and-out *American* call option by using a multiperiod binomial model.

Exercise 13.8 Explain why Equation (13.5) is valid.

REFERENCES

[1] Boyle, P., M. Broadie, and P. Glasserman (1997). "Monte Carlo Methods for Security Pricing." *Journal of Economic Dynamics and Control* 21: 1267–1321.

[2] Conze, A., and R. Viswanathan (1991). "Path Dependent Options: The Case of Lookback Options." *Journal of Finance* 46: 1893–1907.

[3] Goldman, B., H. Sosin, and M. A. Gatto (1979). "Path Dependent Options: Buy at the Low, Sell at the High." *Journal of Finance* 34: 1111–27.

[4] Hull, J. C., and A. White (1998). "The Use of the Control Variate Technique in Option Pricing." *Journal of Financial and Quantitative Analysis* 23: 237–51.

[5] Ross, S. M. (2002). *Simulation*, 3rd ed. Orlando, FL: Academic Press.

[6] Ross, S. M., and J. G. Shanthikumar (2000). "Pricing Exotic Options: Monotonicity in Volatility and Efficient Simulations." *Probability in the Engineering and Informational Sciences* 14: 317–26.

[7] Ross, S. M., and S. Ghamami (2010). "Efficient Monte Carlo Barrier Option Pricing When the Underlying Security Price Follows a Jump-Diffusion Process." *The Journal of Derivatives* 17(3): 45–52.

[8] Rubinstein, M. (1991). "Pay Now, Choose Later." *Risk* (February).

14. Beyond Geometric Brownian Motion Models

14.1 Introduction

As previously noted, a key premise underlying the assumption that the prices of a security over time follow a geometric Brownian motion (and hence underlying the Black–Scholes option price formula) is that future price changes are independent of past price movements. Many investors would agree with this premise, although many others would disagree. Those accepting the premise might argue that it is a consequence of the *efficient market hypothesis,* which claims that the present price of a security encompasses all the presently available information – including past prices – concerning this security. However, critics of this hypothesis argue that new information is absorbed by different investors at different rates; thus, past price movements are a reflection of information that has not yet been universally recognized but *will* affect future prices. It is our belief that there is no a priori reason why future price movements should necessarily be independent of past movements; one should therefore look at real data to see if they are consistent with the geometric Brownian motion model. That is, rather than taking an a priori position, one should let the data decide as much as possible.

In Section 14.2 we analyze the sequence of nearest-month end-of-day prices of crude oil from 3 January 1995 to 19 November 1997 (a period right before the beginning of the Asian financial crisis that deeply affected demand and, as a result, led to lower crude prices). As part of our analysis, we argue that such a price sequence is not consistent with the assumption that crude prices follow a geometric Brownian motion. In Section 14.3 we offer a new model that is consistent with the data as well as intuitively plausible, and we indicate how it may be used to obtain option prices under (a) the assumption that the future resembles the past and (b) a risk-neutral valuation based on the new model.

266 Beyond Geometric Brownian Motion Models

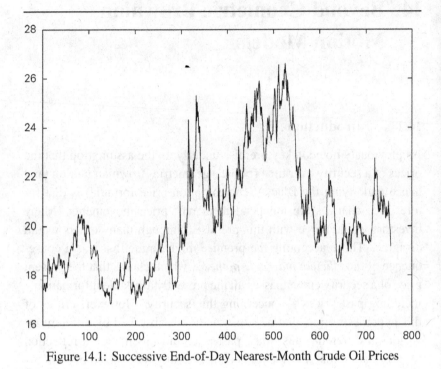

Figure 14.1: Successive End-of-Day Nearest-Month Crude Oil Prices

14.2 Crude Oil Data

With day 0 defined to be 3 January 1995, let $P(n)$ denote the nearest-month price of crude oil (as traded on the New York Mercantile Exchange) at the end of the nth trading day from day 0. The values of $P(n)$ for $n = 1, \ldots, 752$ are given in Figure 14.1 (and in Table 14.5, located at the end of this chapter).

Let

$$L(n) = \log(P(n)),$$

and define

$$D(n) = L(n) - L(n-1).$$

That is, $D(n)$ for $n \geq 1$ are the successive differences in the logarithms of the end-of-day prices. The values of the $D(n)$ are also given in Table 14.5, and Figure 14.2 presents a histogram of those data.

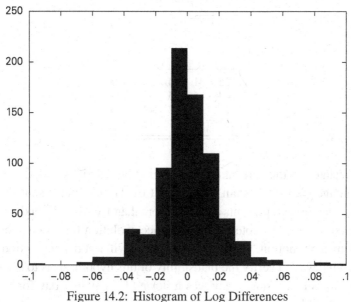

Figure 14.2: Histogram of Log Differences

Note that, under geometric Brownian motion, the $D(n)$ would be independent and identically distributed normal random variables; the histogram in Figure 14.2 is consistent with the hypothesis that the data come from a normal population. However, a histogram – which breaks up the range of data values into intervals and then plots the number of data values that fall in each interval – is not informative about possible dependencies among the data. To consider this possibility, let us classify each day as being in one of four possible *states* as follows: the state of day n is

$$1 \quad \text{if } D(n) \le -.01,$$

$$2 \quad \text{if } -.01 < D(n) \le 0,$$

$$3 \quad \text{if } 0 < D(n) \le .01,$$

$$4 \quad \text{if } D(n) > .01.$$

That is, day n is in state 1 if its end-of-day price represents a loss of more than 1% ($e^{-.01} \approx .99005$) from the end-of-day price on day $n - 1$;

Table 14.1

		j			
i	1	2	3	4	Total
1	55	41	44	36	176
2	44	65	45	60	214
3	26	46	47	49	168
4	52	62	31	48	193

it is in state 2 if the percentage loss is less than 1%; it is in state 3 if the percentage *gain* is less than 1% ($e^{.01} \approx 1.0101$); and it is in state 4 if its end-of-day price represents a gain of more than 1% from the end-of-day price on day $n - 1$. Note that, if the price evolution follows a geometric Brownian motion, then tomorrow's state will not depend on today's state. One way to verify the plausibility of this hypothesis is to see how many times that a state i day was followed by a state j day for $i, j = 1, \ldots, 4$. Table 14.1 gives this information and shows, for instance, that 26 of the 168 days in state 3 were followed by a state-1 day, 46 were followed by a state-2 day, and so on.

The implications of Table 14.1 become clearer if we express the data in terms of percentages, as is done in Table 14.2. Thus, for instance, a large drop (more than 1%) was followed 31% of the time by another large drop, 23% of the time by a small drop, 25% of the time by a small increase, and 21% of the time by a large increase. It is interesting to note that, whereas a moderate gain was followed by a large drop 15% of the time, a large gain was followed by a large drop 27% of the time. Under the geometric Brownian motion model, tomorrow's change would be unaffected by today's change and so the theoretically expected percentages in Table 14.2 would be the same for all rows. To see how likely it is that the actual data would have occurred under geometric Brownian motion, we can employ a standard statistical procedure (testing for independence in a contingency table); using this procedure on our data results in a p-value equal to .005. This means that if the row probabilities were equal (as implied by geometric Brownian motion), then the probability that the resulting data would be as nonsupportive of this hypothesized equality as our actual data is only about 1 in

Table 14.2

		j		
i	1	2	3	4
1	31	23	25	21
2	21	30	21	28
3	15	28	28	29
4	27	32	16	25

200. (The value of the test statistics is 23.447, resulting in a *p*-value of .00526.)

Let us now break up the data, which consists of 751 $D(n)$ values, into four groupings: the first group consists of the 176 values (of the log of tomorrow's price minus the log of today's) for which today's state is 1, and so on with the other groupings. Figures 14.3–14.6 present the histograms of the data values in each group. Note that each histogram has (approximately) the bell-shaped form of the normal density function.

Let $\bar{x}_i$ and s_i be, respectively, the sample mean and sample standard deviation (equal to the square root of the sample variance) of grouping i for $i = 1, 2, 3, 4$. A computation produces the values listed in Table 14.3.

Under the geometric Brownian motion model, the four data sets will all come from the same normal population and hence we could use a standard statistical test – called a one-way analysis of variance – to test the hypothesis that all four data sets describe normal random variables having the same mean and variance. The necessary calculations reveal that the test statistic (which, when the hypothesis is true, has an F distribution with 3 numerator and 747 denominator degrees of freedom) has a value of 4.50, which is quite large. Indeed, if the hypothesis were true then the probability that the test statistic would have a value at least this large is less than .001, giving us additional evidence that the crude oil data does not follow a geometric Brownian motion. (We could also test the hypothesis that the variances – but not necessarily the means – are equal by using Bartlett's test for the equality of variances; using our data, the test statistic has value 9.59 with a resulting *p*-value less than .025.)

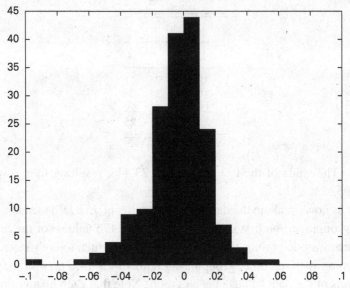

Figure 14.3: Histogram of Post–State-1 Outcomes ($n = 176$)

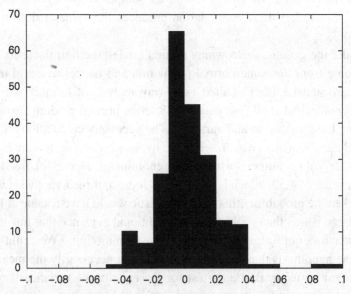

Figure 14.4: Histogram of Post–State-2 Outcomes ($n = 214$)

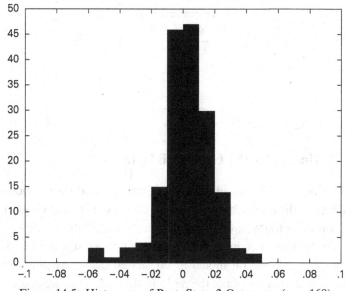

Figure 14.5: Histogram of Post–State-3 Outcomes ($n = 168$)

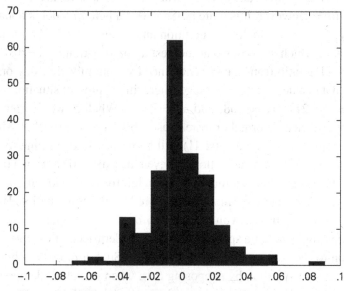

Figure 14.6: Histogram of Post–State-4 Outcomes ($n = 193$)

Table 14.3

i	Mean $\bar{x}_i$	S.D. s_i
1	−.0036	.0194
2	.0024	.0188
3	.0025	.0165
4	−.0011	.0208

14.3   Models for the Crude Oil Data

A reasonable model is to suppose that there are four distributions that determine the difference between the logarithm of tomorrow's price and the logarithm of today's, with the appropriate distribution depending on today's state. However, even within this context we still need to decide if we want a risk-neutral model or one based on the assumption that the future will tend to follow the past. In the latter case we could use a model that supposes, if today's state is i, that the logarithm of the ratio of tomorrow's price to today's price is a normal random variable with mean $\bar{x}_i$ and standard deviation s_i, where these quantities are as given in Table 14.3. However, it is quite possible that a better model is obtained by forgoing the normality assumption and using instead a "bootstrap" approach, which supposes that the best approximation to the distribution of a log ratio from state i is obtained by randomly choosing one of the n_i data values in this grouping (where, in the present situation, $n_1 = 176$, $n_2 = 214$, $n_3 = 168$, and $n_4 = 193$). Whether we assume that the group data are normal or instead use a bootstrap approach, a Monte Carlo simulation (see Chapter 11) will be needed to determine the expected value of owning an option – or even the expected value of a future price. However, such a simulation is straightforward, and variance reduction techniques are available that can reduce the computational time.

A risk-neutral model would appear to be the most appropriate type for assessing whether a specified option is underpriced or overpriced in relation to the present price of the security. Such a model is obtained in the present situation by supposing that, when in state i, the next log ratio is a normal random variable with standard deviation (i.e. volatility) s_i and mean μ_i, where

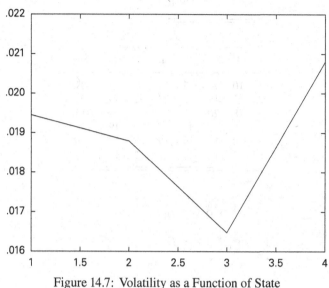

Figure 14.7: Volatility as a Function of State

$$\mu_i = r/N - s_i^2/2;$$

r is the interest rate, and N (usually taken equal to 252) is the number of trading days in a year. Again, a simulation would be needed to determine the expected worth of an option.

Whereas we have chosen to define four different states depending on the ratio of successive end-of-day prices, it is quite possible that a better model could be obtained by allowing for more states. Indeed, one approach for obtaining a risk-neutral model is to assess the volatility as a function of the most recent value of $D(n)$ – by assuming that the volatility is equal to s_i when $D(n)$ is the midpoint of region i – and then to use a general linear interpolation scheme (see Figure 14.7).

Rather than having four different states, we might rather have defined *six* states as follows: the state of day n is

1 if $D(n) \le -.02$,

2 if $-.02 < D(n) \le .01$,

3 if $-.01 < D(n) \le 0$,

Table 14.4

			j				
i	1	2	3	4	5	6	Total
1	10	12	25	19	12	3	81
2	17	16	16	25	12	9	95
3	18	26	65	45	31	29	214
4	11	15	46	47	30	19	168
5	14	15	39	19	13	10	110
6	12	11	23	12	12	13	83

$$4 \quad \text{if } 0 < D(n) \le .01,$$

$$5 \quad \text{if } .01 < D(n) \le .02,$$

$$6 \quad \text{if } D(n) > .02.$$

With these states, the number of times that a state-i day was followed by a state-j day is as given in row i, column j of Table 14.4. The resulting model can then be analyzed in exactly the same manner as was the four-state model.

14.4   Final Comments

We have seen in this chapter that not all security price data is consistent with the assumption that its price history follows a geometric Brownian motion. Geometric Brownian motion is a Markov model, which is one that supposes that a future state of the system (i.e., price of the security) depends only on the present state and not on any previous states. However, to many people it seems reasonable that a security's recent price history can be somewhat useful in predicting future prices. In this chapter we have proposed a simple model for end-of-day prices, one in which the successive ratios of the price on day n to the price on day $n - 1$ are assumed to constitute a Markov model. That is, with regard to the successive ratios of prices, geometric Brownian motion supposes that they are independent whereas our proposed model allows them to have a Markov dependence.

In using the model to value an option, we recommend that one collect up-to-date data and then model the future under the assumption that it will follow the past, either by using a bootstrap approach or by assuming normality and using the estimates $\bar{x}_i$ and s_i. However, if one wants to determine whether an option is underpriced or overpriced in relation to the security itself, we recommend using the risk-neutral variant of the model. This latter model takes $r/N - s_i^2/2$, rather than $\bar{x}_i$, as the mean of a log ratio from state i. This risk-neutral model, which allows the volatility to depend on the most recent daily change, is consistent with a variant of the efficient market hypothesis which states that the present price of a security is the "fair price," in the sense that the expectation of the present value of a future price is equal to the present price (this is known as the *martingale* hypothesis).

REFERENCES

[1] Efron, B., and R. Tibshirani (1993). *An Introduction to the Bootstrap.* New York: Chapman and Hall.
[2] Fama, Eugene (1965). "The Behavior of Stock Market Prices." *Journal of Business* 38: 34–105.
[3] Malkiel, Burton G. (1990). *A Random Walk Down Wall Street.* New York: Norton.
[4] Niederhoffer, Victor (1966). "A New Look at Clustering of Stock Prices." *Journal of Business* 39: 309–13.

Table 14.5: *Nearest-Month Crude Oil Data* (dollars)

Date	Price	Log Difference	Date	Price	Log Difference
1/3/95	17.44		3/7/95	18.63	0.00214938
1/4/95	17.48	0.00229095	3/8/95	18.33	−0.0162341
1/5/95	17.72	0.0136366	3/9/95	18.02	−0.0170568
1/6/95	17.67	−0.00282566	3/10/95	17.91	−0.00612304
1/9/95	17.4	−0.0153981	3/13/95	18.19	0.0155128
1/10/95	17.37	−0.00172563	3/14/95	17.94	−0.0138391
1/11/95	17.72	0.0199494	3/15/95	18.11	0.00943142
1/12/95	17.72	0	3/16/95	18.16	0.0027571
1/13/95	17.52	−0.0113509	3/17/95	18.26	0.0054915
1/16/95	17.88	0.0203397	3/20/95	18.56	0.0162959
1/17/95	18.32	0.0243106	3/21/95	18.43	−0.00702896
1/18/95	18.73	0.0221332	3/22/95	18.96	0.0283517
1/19/95	18.69	−0.00213789	3/23/95	18.92	−0.00211193
1/20/95	18.65	−0.00214248	3/24/95	18.78	−0.00742709
1/23/95	18.1	−0.0299342	3/27/95	19.07	0.0153239
1/24/95	18.39	0.0158951	3/28/95	19.05	−0.00104932
1/25/95	18.39	0	3/29/95	19.22	0.0088843
1/26/95	18.24	−0.00819005	3/30/95	19.15	−0.00364869
1/27/95	17.95	−0.0160269	3/31/95	19.17	0.00104384
1/30/95	18.09	0.00776918	4/3/95	19.03	−0.00732988
1/31/95	18.39	0.0164477	4/4/95	19.18	0.00785139
2/1/95	18.52	0.00704419	4/5/95	19.56	0.0196186
2/2/95	18.54	0.00107933	4/6/95	19.77	0.010679
2/3/95	18.78	0.0128619	4/7/95	19.67	−0.005071
2/6/95	18.59	−0.0101687	4/10/95	19.59	−0.0040754
2/7/95	18.46	−0.00701757	4/11/95	19.88	0.014695
2/8/95	18.3	−0.00870517	4/12/95	19.55	−0.0167389
2/9/95	18.24	−0.00328408	4/13/95	19.15	−0.0206726
2/10/95	18.46	0.0119892	4/17/95	19.15	0
2/13/95	18.27	−0.0103459	4/18/95	19.73	0.0298376
2/14/95	18.32	0.00273299	4/19/95	20.05	0.0160888
2/15/95	18.42	0.00544367	4/20/95	20.41	0.0177958
2/16/95	18.59	0.00918677	4/21/95	20.52	0.00537504
2/17/95	18.91	0.0170671	4/24/95	20.41	−0.00537504
	18.91	0	4/25/95	20.12	−0.0143106
2/21/95	18.86	−0.00264761	4/26/95	20.29	0.00841381
2/22/95	18.63	−0.0122701	4/27/95	20.15	−0.00692387
2/23/95	18.43	−0.0107934	4/28/95	20.43	0.0138001
2/24/95	18.69	0.0140088		20.38	−0.00245038
2/27/95	18.66	−0.00160643	5/1/95	20.5	0.00587086
2/28/95	18.49	−0.00915215	5/2/95	20.09	−0.0202027
3/1/95	18.32	−0.00923669	5/3/95	19.89	−0.0100051
3/2/95	18.35	0.00163622	5/4/95	20.29	0.0199111
3/3/95	18.63	0.0151436	5/5/95	20.33	0.00196947
3/6/95	18.59	−0.00214938	5/8/95	20.29	−0.00196947

Table 14.5 *(cont.)*

Date	Price	Log Difference	Date	Price	Log Difference
5/9/95	19.61	−0.0340885	7/11/95	17.32	−0.00115407
5/10/95	19.75	0.00711385	7/12/95	17.49	0.00976739
5/11/95	19.41	−0.0173651	7/13/95	17.25	−0.0138171
5/12/95	19.52	0.00565118	7/14/95	17.32	0.00404976
5/15/95	19.9	0.0192802	7/17/95	17.2	−0.00695252
5/16/95	20.08	0.00900456	7/18/95	17.35	0.00868312
5/17/95	19.96	−0.00599402	7/19/95	17.33	−0.0011534
5/18/95	20	0.002002	7/20/95	17.01	−0.0186377
5/19/95	20.06	0.00299551	7/21/95	16.79	−0.0130179
5/22/95	19.81	−0.0125409	7/24/95	16.88	0.00534602
5/23/95	19.77	−0.00202122	7/25/95	16.93	0.00295771
5/24/95	19.41	−0.0183772	7/26/95	17.5	0.0331137
5/25/95	19.26	−0.00775799	7/27/95	17.49	−0.000571592
5/26/95	18.69	−0.0300418	7/28/95	17.43	−0.00343643
	18.69	0	7/31/95	17.56	0.00743073
5/30/95	18.78	0.00480385	8/1/95	17.7	0.00794105
5/31/95	18.89	0.00584021	8/2/95	17.78	0.00450959
6/1/95	18.9	0.000529241	8/3/95	17.72	−0.00338028
6/2/95	19.14	0.0126185	8/4/95	17.71	−0.000564493
6/5/95	19.25	0.00573067	8/7/95	17.65	−0.00339367
6/6/95	19.06	−0.00991916	8/8/95	17.79	0.00790072
6/7/95	19.18	0.00627617	8/9/95	17.78	−0.000562272
6/8/95	18.91	−0.0141772	8/10/95	17.89	0.00616767
6/9/95	18.8	−0.00583401	8/11/95	17.86	−0.00167832
6/12/95	18.86	0.00318641	8/14/95	17.48	−0.0215062
6/13/95	18.91	0.00264761	8/15/95	17.47	−0.000572246
6/14/95	19.05	0.00737622	8/16/95	17.55	0.00456883
6/15/95	18.94	−0.00579101	8/17/95	17.66	0.00624825
6/16/95	18.84	−0.00529382	8/18/95	17.87	0.0118211
6/19/95	18.22	−0.0334624	8/21/95	18.25	0.0210418
6/20/95	18.01	−0.0115927	8/22/95	18.54	0.0157655
6/21/95	17.46	−0.0310146	8/23/95	18	−0.0295588
6/22/95	17.5	0.00228833	8/24/95	17.86	−0.00780818
6/23/95	17.49	−0.000571592	8/25/95	17.86	0
6/26/95	17.64	0.00853976	8/28/95	17.82	−0.00224215
6/27/95	17.77	0.00734259	8/29/95	17.82	0
6/28/95	17.97	0.0111921	8/30/95	17.79	−0.00168492
6/29/95	17.56	−0.0230801	8/31/95	17.84	0.00280663
6/30/95	17.4	−0.00915338	9/1/95	18.04	0.0111484
	17.4	0		18.04	0
	17.4	0	9/5/95	18.58	0.0294942
7/5/95	17.18	−0.0127243	9/6/95	18.36	−0.0119113
7/6/95	17.37	0.0109987	9/7/95	18.27	−0.00491401
7/7/95	17.14	−0.0133297	9/8/95	18.44	0.00926185
7/10/95	17.34	0.0116011	9/11/95	18.47	0.00162558

Table 14.5 *(cont.)*

Date	Price	Log Difference	Date	Price	Log Difference
9/12/95	18.64	0.00916202	11/14/95	17.82	0.00112296
9/13/95	18.54	−0.00537925	11/15/95	17.93	0.00615387
9/14/95	18.85	0.0165824	11/16/95	18.19	0.0143967
9/15/95	18.92	0.00370665	11/17/95	18.57	0.0206754
9/18/95	18.93	0.000528402	11/20/95	18.06	−0.0278478
9/19/95	18.95	0.00105597	11/21/95	17.97	−0.00499585
9/20/95	18.69	−0.0138153	11/22/95	17.96	−0.000556638
9/21/95	17.56	−0.062365	11/23/95	17.96	0
9/22/95	17.25	−0.0178114	11/24/95	17.96	0
9/25/95	17.47	0.012673	11/27/95	18.38	0.0231161
9/26/95	17.33	−0.00804602	11/28/95	18.33	−0.00272406
9/27/95	17.57	0.0137538	11/29/95	18.26	−0.00382619
9/28/95	17.76	0.0107558	11/30/95	18.18	−0.00439079
9/29/95	17.54	−0.0124648	12/1/95	18.43	0.0136577
10/2/95	17.64	0.00568506	12/4/95	18.63	0.0107934
10/3/95	17.56	−0.00454546	12/5/95	18.67	0.00214477
10/4/95	17.3	−0.0149171	12/6/95	18.77	0.00534189
10/5/95	16.87	−0.0251696	12/7/95	18.73	−0.00213333
10/6/95	17.03	0.0094396	12/8/95	18.97	0.0127323
10/9/95	17.31	0.0163079	12/11/95	18.66	−0.0164766
10/10/95	17.42	0.0063346	12/12/95	18.73	0.00374432
10/11/95	17.29	−0.00749067	12/13/95	19	0.0143125
10/12/95	17.12	−0.00988093	12/14/95	19.11	0.00577278
10/13/95	17.41	0.0167974	12/15/95	19.51	0.0207154
10/16/95	17.59	0.0102858	12/18/95	19.67	0.00816748
10/17/95	17.68	0.0051035	12/19/95	19.12	−0.0283597
10/18/95	17.61	−0.00396713	12/20/95	18.97	−0.00787612
10/19/95	17.32	−0.016605	12/21/95	18.96	−0.000527287
10/20/95	17.37	0.00288268	12/22/95	19.14	0.00944889
10/23/95	17.21	−0.00925397	12/25/95	19.14	0
10/24/95	17.32	0.00637129	12/26/95	19.27	0.0067691
10/25/95	17.32	0	12/27/95	19.5	0.011865
10/26/95	17.58	0.0149	12/28/95	19.36	−0.00720538
10/27/95	17.54	−0.00227791	12/29/95	19.55	0.0097662
10/30/95	17.62	0.00455063	1/1/96	19.55	0
10/31/95	17.64	0.00113443	1/2/96	19.81	0.0132116
11/1/95	17.74	0.00565293	1/3/96	19.89	0.00403023
11/2/95	17.98	0.0134381	1/4/96	19.91	0.00100503
11/3/95	17.94	−0.00222717	1/5/96	20.26	0.0174264
11/6/95	17.71	−0.0129034	1/8/96	20.26	0
11/7/95	17.65	−0.00339367	1/9/96	19.95	−0.0154194
11/8/95	17.82	0.00958564	1/10/96	19.67	−0.0141345
11/9/95	17.84	0.00112171	1/11/96	18.79	−0.0457698
11/10/95	17.83	−0.000560695	1/12/96	18.25	−0.0291597
11/13/95	17.8	−0.00168397	1/15/96	18.38	0.00709804

Table 14.5 *(cont.)*

Date	Price	Log Difference	Date	Price	Log Difference
1/16/96	18.05	−0.0181174	3/19/96	24.34	0.0449561
1/17/96	18.52	0.0257055	3/20/96	23.06	−0.0540216
1/18/96	19.18	0.0350168	3/21/96	21.05	−0.091199
1/19/96	18.94	−0.012592	3/22/96	21.95	0.0418666
1/22/96	18.62	−0.0170398	3/25/96	22.4	0.0202938
1/23/96	18.06	−0.0305367	3/26/96	22.19	−0.00941922
1/24/96	18.28	0.012108	3/27/96	21.79	−0.0181906
1/25/96	17.67	−0.0339393	3/28/96	21.41	−0.017593
1/26/96	17.73	0.00338983	3/29/96	21.47	0.00279851
1/29/96	17.45	−0.0159185	4/1/96	22.26	0.0361347
1/30/96	17.56	0.00628394	4/2/96	22.7	0.0195736
1/31/96	17.74	0.0101984	4/3/96	22.27	−0.0191244
2/1/96	17.71	−0.00169253	4/4/96	22.75	0.0213247
2/2/96	17.8	0.00506901	4/5/96	22.75	0
2/5/96	17.54	−0.0147145	4/8/96	23.03	0.0122326
2/6/96	17.69	0.00851552	4/9/96	23.06	0.0013018
2/7/96	17.74	0.00282247	4/10/96	24.21	0.0486663
2/8/96	17.76	0.00112676	4/11/96	25.34	0.0456184
2/9/96	17.78	0.00112549	4/12/96	24.29	−0.0423194
2/12/96	17.97	0.0106295	4/15/96	25.06	0.0312082
2/13/96	18.91	0.0509872	4/16/96	24.47	−0.0238251
2/14/96	18.96	0.00264061	4/17/96	24.67	0.00814005
2/15/96	19.04	0.00421053	4/18/96	23.82	−0.0350624
2/16/95	19.16	0.00628274	4/19/96	23.95	0.00544276
2/19/96	19.16	0	4/22/96	24.07	0.00499793
2/20/96	21.05	0.0940758	4/23/96	22.7	−0.0586013
2/21/96	19.71	−0.0657744	4/24/96	22.4	−0.013304
2/22/96	19.85	0.00707789	4/25/96	22.2	−0.00896867
2/23/96	19.06	−0.0406121	4/26/96	22.32	0.00539085
2/26/96	19.39	0.0171656	4/29/96	22.43	0.00491621
2/27/96	19.7	0.0158612	4/30/96	21.2	−0.0563982
2/28/96	19.29	−0.0210318	5/1/96	20.81	−0.0185675
2/29/96	19.54	0.0128768	5/2/96	20.86	0.00239981
3/1/96	19.44	−0.00513085	5/3/96	21.18	0.0152239
3/4/96	19.2	−0.0124225	5/6/96	21.04	−0.00663195
3/5/96	19.54	0.0175534	5/7/96	21.11	0.00332147
3/6/96	20.19	0.0327238	5/8/96	21	−0.00522442
3/7/96	19.81	−0.0190006	5/9/96	20.68	−0.0153554
3/8/96	19.61	−0.0101472	5/10/96	21.01	0.0158315
3/11/96	19.91	0.0151825	5/13/96	21.36	0.0165215
3/12/96	20.46	0.0272496	5/14/96	21.42	0.00280505
3/13/96	20.58	0.00584797	5/15/96	21.48	0.0027972
3/14/96	21.16	0.0277929	5/16/96	20.78	−0.0331313
3/15/96	21.99	0.0384752	5/17/96	20.64	−0.00676005
3/18/96	23.27	0.0565772	5/20/96	22.48	0.0853951

Table 14.5 *(cont.)*

Date	Price	Log Difference	Date	Price	Log Difference
5/21/96	22.65	0.00753383	7/23/96	21.01	−0.0183924
5/22/96	21.4	−0.0567689	7/24/96	20.68	−0.0158315
5/23/96	21.23	−0.00797565	7/25/96	20.74	0.00289715
5/24/96	21.32	0.00423032	7/26/96	20.11	−0.030847
	21.32	0	7/29/86	20.28	0.00841797
5/28/96	21.11	−0.00989874	7/30/96	20.33	0.00246245
5/29/96	20.76	−0.0167188	7/31/96	20.42	0.00441719
5/30/96	19.94	−0.0403003	8/1/96	21.04	0.0299106
5/31/96	19.76	−0.00906807	8/2/96	21.34	0.0141579
6/3/96	19.85	0.00454431	8/5/96	21.23	−0.00516797
6/4/96	20.44	0.0292898	8/6/96	21.13	−0.00472144
6/5/96	19.72	−0.0358604	8/7/96	21.42	0.0136312
6/6/96	20.05	0.0165958	8/8/96	21.55	0.00605075
6/7/96	20.28	0.011406	8/9/96	21.57	0.000927644
6/10/96	20.25	−0.00148039	8/12/96	22.22	0.0296893
6/11/96	20.1	−0.00743498	8/13/96	22.37	0.00672799
6/12/96	20.09	−0.000497636	8/14/96	22.12	−0.0112386
6/13/96	20.01	−0.00399003	8/15/96	21.9	−0.00999554
6/14/96	20.34	0.0163572	8/16/96	22.66	0.0341146
6/17/96	22.14	0.0847965	8/19/96	23.26	0.0261339
6/18/96	21.46	−0.0311952	8/20/96	22.86	−0.0173465
6/19/96	20.76	−0.0331627	8/21/96	21.72	−0.0511552
6/20/96	20.65	−0.00531274	8/22/96	22.3	0.0263532
6/21/96	19.92	−0.0359911	8/23/96	21.96	−0.0153641
6/24/96	19.98	0.00300752	8/26/96	21.62	−0.0156038
6/25/96	19.96	−0.0010015	8/27/96	21.56	−0.00277907
6/26/96	20.65	0.033985	8/28/96	21.71	0.00693324
6/27/96	21.02	0.017759	8/29/96	22.15	0.0200645
6/28/96	20.92	−0.00476873	8/30/96	22.25	0.00450451
7/1/96	21.53	0.0287417	9/2/96	22.25	0
7/2/96	21.13	−0.0187535	9/3/96	23.4	0.050394
7/3/96	21.21	0.00377894	9/4/96	23.24	−0.00686109
7/4/96	21.21	0	9/5/96	23.44	0.00856903
7/5/96	21.21	0	9/6/96	23.85	0.0173403
7/8/96	21.27	0.00282486	9/9/96	23.73	−0.00504415
7/9/96	21.41	0.00656047	9/10/96	24.12	0.0163013
7/10/96	21.55	0.00651771	9/11/96	24.75	0.0257841
7/11/96	21.95	0.0183913	9/12/96	25	0.0100503
7/12/96	21.9	−0.0022805	9/13/96	24.51	−0.0197946
7/15/96	22.48	0.0261394	9/16/96	23.19	−0.05536
7/16/96	22.38	−0.00445832	9/17/96	23.31	0.0051613
7/17/96	21.8	−0.0262577	9/18/96	23.89	0.0245775
7/18/96	21.68	−0.00551979	9/19/96	23.54	−0.0147589
7/19/96	21	−0.0318677	9/20/96	23.63	0.00381599
7/22/96	21.4	0.0188685	9/23/96	23.37	−0.0110639

Table 14.5 *(cont.)*

Date	Price	Log Difference	Date	Price	Log Difference
9/24/96	24.07	0.0295131	11/26/96	23.62	0.00551901
9/25/96	24.46	0.0160729	11/27/96	23.75	0.00548872
9/26/96	24.16	−0.0123408	11/28/96	23.75	0
9/27/96	24.6	0.0180481	11/29/96	23.75	0
9/30/96	24.38	−0.00898322	12/2/96	24.8	0.0432611
10/1/96	24.14	−0.00989291	12/3/96	24.93	0.00522824
10/2/96	24.05	−0.00373522	12/4/96	24.8	−0.00522824
10/3/96	24.81	0.0311118	12/5/96	25.58	0.0309671
10/4/96	24.73	−0.00322972		25.62	0.0015625
10/7/96	25.24	0.020413	12/9/96	25.3	−0.0125689
10/8/96	25.54	0.0118158	12/10/96	24.42	−0.0354019
10/9/96	25.07	−0.0185739	12/11/96	23.38	−0.0435215
10/10/96	24.26	−0.032843	12/12/96	23.72	0.0144376
10/11/96	24.66	0.0163536	12/13/96	24.47	0.0311293
10/14/96	25.62	0.0381908	12/16/96	25.74	0.0505983
10/15/98	25.42	−0.00783703	12/17/96	25.71	−0.00116618
10/16/96	25.17	−0.00988346	12/18/96	26.16	0.0173515
10/17/96	25.42	0.00988346	12/19/96	26.57	0.0155512
10/18/96	25.75	0.0128984	12/20/96	25.08	−0.057712
10/21/96	25.92	0.00658024	12/23/96	24.79	−0.0116304
10/22/96	25.75	−0.00658024	12/24/96	25.1	0.0124275
10/23/96	24.86	−0.0351745	12/25/96	25.1	0
10/24/96	24.51	−0.0141789	12/26/96	24.92	−0.00719715
10/25/96	24.86	0.0141789	12/27/96	25.22	0.0119666
10/28/96	24.85	−0.000402334	12/30/96	25.37	0.00593004
10/29/96	24.34	−0.0207367	12/31/96	25.92	0.0214475
10/30/96	24.28	−0.00246812		25.92	0
10/31/96	23.35	−0.039056	1/2/97	25.69	−0.00891306
11/1/96	23.03	−0.0137993	1/3/97	25.59	−0.00390016
11/4/96	22.79	−0.0104759	1/6/97	26.37	0.0300254
11/5/96	22.64	−0.00660359	1/7/97	26.23	−0.00532321
11/6/96	22.69	0.00220605	1/8/97	26.62	0.014759
11/7/96	22.74	0.00220119	1/9/97	26.37	−0.00943581
11/8/96	23.59	0.0366974	1/10/97	26.09	−0.0106749
11/11/96	23.37	−0.00936974	1/13/97	25.19	−0.035105
11/12/96	23.35	−0.000856164	1/14/97	25.11	−0.00318092
11/13/96	24.12	0.0324444	1/15/97	25.95	0.0329054
11/14/96	24.41	0.0119515	1/16/97	25.52	−0.0167072
11/15/96	24.17	−0.00988069	1/17/97	25.41	−0.00431966
11/18/96	23.88	−0.0120709	1/20/97	25.23	−0.00710903
11/19/96	24.49	0.0252236	1/21/97	24.8	−0.0171901
11/20/96	23.76	−0.0302614	1/22/97	24.24	−0.0228395
11/21/96	23.84	0.00336135	1/23/97	24.18	−0.00247832
11/22/96	23.75	−0.00378231	1/24/97	24.05	−0.00539085
11/25/96	23.49	−0.0110077	1/27/97	23.94	−0.0045843

Table 14.5 *(cont.)*

Date	Price	Log Difference	Date	Price	Log Difference
1/28/97	23.9	−0.00167224	4/1/97	20.28	−0.0063898
1/29/97	24.47	0.0235694	4/2/97	19.47	−0.0407604
1/30/97	24.87	0.0162144	4/3/97	19.47	0
1/31/97	24.15	−0.0293779	4/4/97	19.12	−0.0181399
2/3/97	24.15	0	4/7/97	19.23	0.00573665
2/4/97	24.02	−0.00539756	4/8/97	19.35	0.00622086
2/5/97	23.91	−0.00459004	4/9/97	19.27	−0.00414294
2/6/97	23.1	−0.0344642	4/10/97	19.57	0.0154483
2/7/97	22.23	−0.0383899	4/11/97	19.53	−0.00204604
2/10/97	22.46	0.0102932	4/14/97	19.9	0.018768
2/11/97	22.42	−0.00178253	4/15/97	19.83	−0.00352379
2/12/97	21.86	−0.0252949	4/16/97	19.35	−0.0245035
2/13/97	22.02	0.00729265	4/17/97	19.42	0.00361104
2/14/97	22.41	0.0175562	4/18/97	19.91	0.0249187
2/17/97	22.41	0	4/21/97	20.38	0.0233319
2/18/97	22.52	0.00489652	4/22/97	19.6	−0.0390245
2/19/97	22.79	0.011918	4/23/97	19.73	0.00661075
2/20/97	21.98	−0.0361889	4/24/97	20.03	0.0150908
2/21/97	21.39	−0.0272094	4/25/97	19.99	−0.001999
2/24/97	20.71	−0.0323068	4/28/97	19.91	−0.00401003
2/25/97	21	0.0139058	4/29/97	20.44	0.0262716
2/26/97	21.11	0.00522442	4/30/97	20.21	−0.0113162
2/27/97	20.89	−0.0104763	5/1/97	19.91	−0.0149554
2/28/97	20.3	−0.0286497	5/2/97	19.6	−0.0156926
3/3/97	20.25	−0.00246609	5/5/97	19.63	0.00152944
3/4/97	20.66	0.0200447	5/6/97	19.66	0.00152711
3/5/97	20.49	−0.0082625	5/7/97	19.62	−0.00203666
3/6/97	20.94	0.0217242	5/8/97	20.34	0.0360399
3/7/97	21.28	0.0161065	5/9/97	20.43	0.00441502
3/10/97	20.49	−0.0378307	5/12/97	21.38	0.0454515
3/11/97	20.11	−0.0187198	5/13/97	21.37	−0.000467836
3/12/97	20.62	0.0250443	5/14/97	21.39	0.000935454
3/13/97	20.7	0.00387222	5/15/97	21.3	−0.00421645
3/14/97	21.29	0.0281038	5/16/97	22.12	0.0377751
3/17/97	20.92	−0.0175318	5/19/97	21.59	−0.0242519
3/18/97	22.06	0.0530604	5/20/97	21.19	−0.0187009
3/19/97	22.04	−0.00090703	5/21/97	21.86	0.0311291
3/20/97	22.32	0.0126242	5/22/97	21.86	0
3/21/97	21.51	−0.0369652	5/23/97	21.63	−0.0105772
3/24/97	21.06	−0.0211424	5/26/97	21.63	0
3/25/97	20.99	−0.00332937	5/27/97	20.79	−0.0396091
3/26/97	20.64	−0.0168152	5/28/97	20.79	0
3/27/97	20.7	0.00290276	5/29/97	20.97	0.00862074
3/28/97	20.7	0	5/30/97	20.88	−0.00430108
3/31/97	20.41	−0.0141087	6/2/97	21.12	0.0114287

Table 14.5 *(cont.)*

Date	Price	Log Difference	Date	Price	Log Difference
6/3/97	20.33	−0.0381228	8/5/97	20.81	0.00288739
6/4/97	20.12	−0.0103833	8/6/97	20.46	−0.0169619
6/5/97	19.66	−0.0231282	8/7/97	20.09	−0.0182496
6/6/97	18.79	−0.0452613	8/8/97	19.54	−0.0277585
6/9/97	18.68	−0.00587138	8/11/97	19.69	0.00764725
6/10/97	18.67	−0.000535475	8/12/97	19.99	0.0151213
6/11/97	18.53	−0.00752692	8/13/97	20.19	0.00995528
6/12/97	18.69	0.00859758	8/14/97	20.08	−0.00546314
6/13/97	18.83	0.00746272	8/15/97	20.07	−0.000498132
6/16/97	19.01	0.00951381	8/18/97	19.91	−0.00800404
6/17/97	19.23	0.0115064	8/19/97	20.12	0.0104922
6/18/97	18.79	−0.0231467	8/20/97	20.06	−0.00298656
6/19/97	18.67	−0.00640686	8/21/97	19.66	−0.0201417
6/20/97	18.55	−0.00644817	8/22/97	19.7	0.00203252
6/23/97	19.14	0.0313106	8/25/97	19.26	−0.0225882
6/24/97	19.03	−0.0057637	8/26/97	19.28	0.00103788
6/25/97	19.52	0.0254229	8/27/97	19.73	0.023072
6/26/97	19.09	−0.0222749	8/28/97	19.58	−0.00763168
6/27/97	19.46	0.0191964	8/29/97	19.61	0.001531
6/30/97	19.8	0.0173209	9/1/97	19.61	0
7/1/97	20.12	0.0160324	9/2/97	19.65	0.0020377
7/2/97	20.34	0.010875	9/3/97	19.61	−0.0020377
7/3/97	19.56	−0.0391027	9/4/97	19.4	−0.0107666
7/4/97	19.56	0	9/5/97	19.63	0.0117859
7/7/97	19.52	−0.00204708	9/8/97	19.45	−0.00921194
7/8/97	19.73	0.0107007	9/9/97	19.42	−0.00154361
7/9/97	19.46	−0.0137792	9/10/97	19.42	0
7/10/97	19.22	−0.0124097	9/11/97	19.37	−0.00257799
7/11/97	19.33	0.00570689	9/12/97	19.32	−0.00258465
7/14/97	18.99	−0.0177458	9/15/97	19.27	−0.00259135
7/15/97	19.67	0.0351821	9/16/97	19.61	0.0174902
7/16/97	19.65	−0.00101729	9/17/97	19.42	−0.00973618
7/17/97	19.99	0.0171548	9/18/97	19.38	−0.00206186
7/18/97	19.27	−0.0366827	9/19/97	19.35	−0.00154919
7/21/97	19.18	−0.00468141	9/22/97	19.6	0.0128371
7/22/97	19.08	−0.0052274	9/23/97	19.79	0.00964719
7/23/97	19.63	0.0284183	9/24/97	19.94	0.007551
7/24/97	19.77	0.00710663	9/25/97	20.39	0.0223168
7/25/97	19.89	0.00605146	9/26/97	20.87	0.0232681
7/28/97	19.81	−0.00403023	9/29/97	21.26	0.0185147
7/29/97	19.85	0.00201715	9/30/97	21.18	−0.00377003
7/30/97	20.3	0.0224169	10/1/97	21.05	−0.00615678
7/31/97	20.14	−0.007913	10/2/97	21.77	0.0336323
8/1/97	20.28	0.00692729	10/3/97	22.76	0.0444717
8/4/97	20.75	0.0229111	10/6/97	21.93	−0.037149

Table 14.5 *(cont.)*

Date	Price	Log Difference	Date	Price	Log Difference
10/7/97	21.96	0.00136705	10/29/97	20.71	0.0121449
10/8/97	22.18	0.00996837	10/30/97	21.22	0.0243275
10/9/97	22.12	−0.00270881	10/31/97	21.08	−0.00661941
10/10/97	22.1	−0.000904568	11/3/97	20.96	−0.00570886
10/13/97	21.32	−0.035932	11/4/97	20.7	−0.0124822
10/14/97	20.7	−0.0295119	11/5/97	20.31	−0.0190203
10/15/97	20.57	−0.0063	11/6/97	20.39	0.00393121
10/16/97	20.97	0.0192591	11/7/97	20.77	0.0184651
10/17/97	20.59	−0.0182873	11/10/97	20.4	−0.0179747
10/20/97	20.7	0.00532818	11/11/97	20.51	0.00537767
10/21/97	20.67	−0.00145033	11/12/97	20.49	−0.00097561
10/22/97	21.42	0.0356417	11/13/97	20.7	0.0101967
10/23/97	21.09	−0.0155261	11/14/97	21	0.0143887
10/24/97	20.97	−0.00570615	11/17/97	20.26	−0.0358739
10/27/97	21.07	0.00475738	11/18/97	20.04	−0.0109182
10/28/97	20.46	−0.0293785	11/19/97	19.8	−0.0120483

15. Autoregressive Models and Mean Reversion

15.1 The Autoregressive Model

Let $S_d(n)$ be the price of a security at the end of day n. If we also let

$$L(n) = \log(S_d(n)),$$

then the geometric Brownian motion model implies that

$$L(n) = a + L(n-1) + e(n), \qquad (15.1)$$

where $e(n)$, $n \geq 1$, is a sequence of independent and identically distributed normal random variables with mean 0 and variance σ^2/N (with $N = 252$ as the number of trading days in a year) and a is equal to μ/N. As before, μ is the mean (or drift) parameter of the geometric Brownian motion and σ is the associated volatility parameter.

Looking at Equation (15.1), it is natural to consider fitting a more general equation for $L(n)$; namely, the linear regression equation

$$L(n) = a + bL(n-1) + e(n), \qquad (15.2)$$

where b is another constant whose value would need to be estimated. That is, rather than arbitrarily taking $b = 1$, an improved model might be obtained by letting b's value be determined by data. Equation (15.2) is the classical linear regression model, and the technique for estimating a, b, and σ is well known. Because the linear regression model given by Equation (15.2) specifies the log price at time n in terms of the log price one time period earlier, it is called an *autoregressive model* of order 1.

The parameters a and b of the autoregressive model given by (15.2) are estimated from historical data in the following manner. Suppose $L(0)$, $L(1), \ldots, L(r)$ are the logarithms of the end-of-day prices for r successive days. Then, when a and b are known, the predicted value of $L(i)$ based on prior log prices is $a + bL(i-1)$; hence, the usual approach to estimating a and b is to let them be the values that minimize the sum of squares of the prediction errors. That is, a and b are chosen to minimize

$$\sum_{i=1}^{r} (L(i) - a - bL(i-1))^2.$$

There are many standard statistical software packages that can be used to calculate the minimimizing values and also to estimate σ.

Remark. The model specified by Equation (15.2) is a risk-neutral model only when $a = (r - \sigma^2/2)/N$ and $b = 1$. That is, it is risk-neutral only when it reduces to the risk-neutral geometric Brownian motion model. Consequently, no arbitrage is possible when all investments are priced according to their expected present values when $a = (r - \sigma^2/2)/N$ and $b = 1$. However, an investor who believes that a and b have some other values can often make an investment that, although not yielding a sure win, can generate a return with a large expected value and a small variance when these latter quantities are computed according to the investor's estimated values of a and b.

15.2   Valuing Options by Their Expected Return

Assume that the end-of-day log prices follow Equation (15.2) and that the parameters a, b, σ have been determined, and consider an option whose exercise time is at the end of n trading days. In order to assess the expected value of this option's payoff, we must first determine the probability distribution of $L(n)$. To accomplish this, start by rewriting the Equation (15.2) as

$$L(i) = e(i) + a + bL(i-1).$$

Now, continually using the preceding equation – first with $i = n$, then with $i = n - 1$, and so on – yields

$$L(n) = e(n) + a + bL(n-1)$$

$$= e(n) + a + b[e(n-1) + a + bL(n-2)]$$

$$= e(n) + be(n-1) + a + ab + b^2 L(n-2)$$

$$= e(n) + be(n-1) + a + ab + b^2 [e(n-2) + a + bL(n-3)]$$

$$= e(n) + be(n-1) + b^2 e(n-2)$$

$$+ a + ab + ab^2 + b^3 L(n-3).$$

Continuing on in this fashion shows that, for any $k < n$,.

$$L(n) = \sum_{i=0}^{k} b^i e(n-i) + a \sum_{i=0}^{k} b^i + b^{k+1} L(n-k-1).$$

Hence, with $k = n - 1$, the preceding equation yields

$$L(n) = \sum_{i=0}^{n-1} b^i e(n-i) + a \sum_{i=0}^{n-1} b^i + b^n L(0)$$

$$= \sum_{i=0}^{n-1} b^i e(n-i) + \frac{a(1-b^n)}{1-b} + b^n L(0). \tag{15.3}$$

Note that $b^i e(n-i)$ is a normal random variable with mean 0 and variance $b^{2i} \sigma^2 / N$. Thus – using that the sum of independent normal random variables is also a normal random variable – we see that $\sum_{i=0}^{n-1} b^i e(n-i)$ is a normal random variable with mean

$$E\left[\sum_{i=0}^{n-1} b^i e(n-i)\right] = \sum_{i=0}^{n-1} b^i E[e(n-i)] = 0 \tag{15.4}$$

and variance

$$\text{Var}\left[\sum_{i=0}^{n-1} b^i e(n-i)\right] = \sum_{i=0}^{n-1} \text{Var}[b^i e(n-i)]$$

$$= \frac{\sigma^2}{N} \sum_{i=0}^{n-1} b^{2i}$$

$$= \frac{\sigma^2(1-b^{2n})}{N(1-b^2)}. \tag{15.5}$$

Hence, from Equations (15.3), (15.4), and (15.5) we obtain that if the logarithm of the price at time 0 is $L(0) = g$, then $L(n)$ is a normal random variable with mean $m(n)$ and variance $v(n)$, where

$$m(n) = \frac{a(1-b^n)}{1-b} + b^n g \tag{15.6}$$

and

$$v(n) = \frac{\sigma^2(1-b^{2n})}{N(1-b^2)}. \tag{15.7}$$

The present value of the payoff of a call option (whose strike price is K and whose exercise time is at the end of n trading days) is

$$e^{-rn/N}(S_d(n) - K)^+ = e^{-rn/N}(e^{L(n)} - K)^+,$$

where r and N are (respectively) the interest rate and the number of trading days in a year. Using that $L(n)$ is normal with mean and variance as given by Equations (15.6) and (15.7), it can be shown that the expected value of this payoff is

$$E[e^{-rn/N}(e^{L(n)} - K)^+]$$
$$= e^{-rn/N}\left(e^{m(n)+v(n)/2}\Phi(\sqrt{v(n)} - h) - K\Phi(-h)\right), \quad (15.8)$$

where Φ is the standard normal distribution function and where

$$h = \frac{\log(K) - m(n)}{\sqrt{v(n)}}.$$

Example 15.2a   Assuming that an autoregressive model is appropriate for the crude oil data from Chapter 12, the estimates of a, b, and $\sigma/\sqrt{N}$ obtained from a standard statistical package are

$$a = .0487, \quad b = .9838, \quad \sigma/\sqrt{N} = .01908.$$

That is, the estimated autoregressive equation is

$$L(n) = .0487 + .9838L(n - 1) + e(n),$$

where $e(n)$ is a normal random variable having mean 0 and standard deviation .01908. Consequently, if the present price is 20, then the logarithm of the price at the end of another 50 trading days is a normal random variable with mean

$$m(50) = \frac{.0487(1 - .9838^{50})}{1 - .9838} + \log(20)(.9838)^{50} = 3.0016$$

and variance

$$v(50) = (.0191)^2\frac{1 - (.9838)^{100}}{1 - (.9838)^2} = .0091.$$

Suppose now that the interest rate is 8% and that we want to determine the expected present value of the payoff from an option to purchase the security at the end of 50 trading days at a strike price $K = 21$. Because

$$h = \frac{\log(21) - 3.0016}{\sqrt{.0091}} = .4499,$$

it follows from Equation (15.8) that the present value of the expected payoff is

$$e^{-.08(50)/252}(20.2094\Phi(-.3545) - 21\Phi(-.4499)) = .4442.$$

That is, the expected present value payoff is 44.42 cents.

It is interesting to compare the preceding result with the geometric Brownian motion Black–Scholes option cost. Using the notation of Section 7.2, the data set of the crude oil prices results in the following estimate of the volatility parameter:

$$\sigma = .3032 \quad (\sigma/\sqrt{N} = .01910).$$

As this gives $\omega = -.1762$ and $\sigma\sqrt{t} = .1351$, the Black–Scholes cost is

$$C = 20\Phi(-.1762) - 21e^{-4/252}\Phi(-.3113) = .7911.$$

Thus the geometric Brownian motion risk-neutral cost valuation of 79 cents is quite a bit more than the expected present value payoff of 44 cents when the autoregressive model is assumed. The primary reason for this discrepancy is that the variance of the logarithm of the final price is .01824 under the risk-neutral geometric Brownian motion model but only .0091 under the autoregresssive model. (The means of the logarithms of the price at exercise time are roughly equal: 3.0025 under the risk-neutral geometric Brownian motion model and 3.0016 under the autoregressive model.)

For additional comparisons, a simulation study yielded that the expected present value of the option payoff under the model of Chapter 12 is 64 cents when the sample means are used as estimators of the mean drifts versus 81 cents when the risk-neutral means are used. □

15.3 Mean Reversion

Many traders believe that the prices of certain securities (often commodities) tend to revert to fixed values. That is, when the current price is less than this value, the price tends to increase; when it is greater, it tends to decrease. Although this phenomenon – called *mean reversion* – cannot be explained by a geometric Brownian motion model, it is a very simple consequence of the autoregressive model. For consider the model

$$L(n) = a + bL(n-1) + e(n),$$

which is equivalent to

$$S_d(n) = e^{a+e(n)}(S_d(n-1))^b.$$

Since

$$E[e^{a+e(n)}] = e^{a+\sigma^2/2N}$$

it follows that, if the price of the security at the end of day $n-1$ is s, then the expected price of the security at the end of the next day is

$$E[S_d(n)] = e^{a+\sigma^2/2N}s^b. \tag{15.9}$$

Now suppose that $0 < b < 1$, and let

$$s^* = \exp\left\{\frac{a+\sigma^2/2N}{1-b}\right\}.$$

We will show that if the present price is s then the expected price at the end of the next day is between s and s^*.

Toward this end, first suppose that $s < s^*$. That is,

$$s < \exp\left\{\frac{a+\sigma^2/2N}{1-b}\right\}, \tag{15.10}$$

which implies that

$$s^{1-b} < \exp\{a+\sigma^2/2N\}$$

or

$$s < \exp\{a+\sigma^2/2N\}s^b = E[S_d(n)]. \tag{15.11}$$

Moreover, Equation (15.10) also implies that

$$s^b < \exp\left\{\frac{b(a+\sigma^2/2N)}{1-b}\right\}$$

or

$$s^b < \exp\left\{\frac{a+\sigma^2/2N}{1-b} - (a+\sigma^2/2N)\right\},$$

which is equivalent to

$$E[S_d(n)] = \exp\{a + \sigma^2/2N\}s^b < \exp\left\{\frac{a + \sigma^2/2N}{1 - b}\right\} = s^*. \quad (15.12)$$

Consequently, from (15.11) and (15.12) we see that, if $S_d(n - 1) = s < s^*$, then

$$s < E[S_d(n)] < s^*.$$

In a similar manner, it follows that if $S_d(n - 1) = s > s^*$ then

$$s^* < E[S_d(n)] < s.$$

Therefore, if $0 < b < 1$ then, for any current end-of-day price s, the mean price at the end of the next day is between s and s^*. In other words, there is a mean reversion to the price s^*.

Example 15.3a For the data of Example 15.2a, the estimated regression equation is

$$L(n) = .0487 + .9838L(n - 1) + e(n),$$

where $e(n)$ is a normal random variable having mean 0 and standard deviation .0191. Since the estimated value of b is less than 1, this model predicts a mean price reversion to the value

$$s^* = \exp\left\{\frac{.0487 + (.0191)^2/2}{1 - .9838}\right\} = 20.44. \qquad \square$$

15.4 Exercises

Exercise 15.1 For the model

$$L(n) = 5 + .8L(n - 1) + e(n),$$

where $e(n)$ is a normal random variable with mean 0 and variance .2, find the probability that $L(n + 10) > L(n)$.

Exercise 15.2 Let $L(n)$ denote the logarithm of the price of a security at the end of day n, and suppose that

$$L(n) = 1.2 + .7L(n-1) + e(n),$$

where $e(n)$ is a normal random variable with mean 0 and variance .1. Find the expected present value payoff of a call option that expires in 60 trading days and has strike price 50 when the interest rate is 10% and the present price of the security is: (a) 48; (b) 50; (c) 52.

Exercise 15.3 Use a statistical package on the first 100 data values for heating oil (presented in Table 15.1, pp. 241–249) to fit an autoregressive model.

Exercise 15.4 To what value does the expected price of the security in Exercise 15.2 revert?

Exercise 15.5 For the model of Section 15.3, show that if $S_d(n-1) = s > s^*$ then

$$s^* < E[S_d(n)] < s.$$

Exercise 15.6 For the model of Section 15.3, show that if $S_d(n-1) = s^*$ then

$$E[S_d(n)] = s^*.$$

Table 15.1: *Nearest-Month Commodity Prices* (dollars)

Date	Unleaded Gas	Heating Oil	Date	Unleaded Gas	Heating Oil
03-Jan-95	52.75	49.94	07-Mar-95	56.78	46.36
04-Jan-95	53.43	49.64	08-Mar-95	55.83	45.25
05-Jan-95	54.51	49.96	09-Mar-95	54.35	45.14
06-Jan-95	53.77	49.52	10-Mar-95	52.47	45.25
09-Jan-95	53.9	48.33	13-Mar-95	53.81	45.61
10-Jan-95	53.66	47.38	14-Mar-95	52.79	44.34
11-Jan-95	54.54	47.98	15-Mar-95	54.04	45.14
12-Jan-95	54.92	47.85	16-Mar-95	54.93	45.37
13-Jan-95	55	46.68	17-Mar-95	55.37	46.07
16-Jan-95	56.88	47.35	20-Mar-95	56.15	45.85
17-Jan-95	57.8	48.67	21-Mar-95	56.15	45.65
18-Jan-95	59.48	49.08	22-Mar-95	55.9	47.02
19-Jan-95	58.12	48.28	23-Mar-95	57.53	46.56
20-Jan-95	57.4	48.14	24-Mar-95	57.82	46.32
23-Jan-95	56.38	47.82	27-Mar-95	58.6	47.46
24-Jan-95	57.6	47.87	28-Mar-95	58.73	47.46
25-Jan-95	57.25	47.47	29-Mar-95	59.99	47.08
26-Jan-95	57.44	47.27	30-Mar-95	60.68	47.19
27-Jan-95	56.07	47.27	31-Mar-95	59.47	47.06
30-Jan-95	56.21	47.42	03-Apr-95	57.44	47.47
31-Jan-95	57.76	46.86	04-Apr-95	58.6	47.96
01-Feb-95	56.77	47.8	05-Apr-95	60.48	48.01
02-Feb-95	55.95	48.55	06-Apr-95	61.68	49.21
03-Feb-95	57.35	49.44	07-Apr-95	61.29	49.5
06-Feb-95	57.3	49.2	10-Apr-95	61.22	49.28
07-Feb-95	56.99	49.13	11-Apr-95	61.59	50.15
08-Feb-95	56.1	47.98	12-Apr-95	61.37	49.54
09-Feb-95	55.84	47.65	13-Apr-95	60.44	48.79
10-Feb-95	55.64	48.28	14-Apr-95	60.44	48.79
13-Feb-95	55.56	47.29	17-Apr-95	62.03	50.01
14-Feb-95	56.16	47.5	18-Apr-95	63.69	50.19
15-Feb-95	56.22	46.89	19-Apr-95	63.15	50.15
16-Feb-95	57.91	46.92	20-Apr-95	63.22	50.28
17-Feb-95	58.76	47.72	21-Apr-95	63.2	50.64
20-Feb-95	58.76	47.72	24-Apr-95	62.21	50.02
21-Feb-95	59.11	47.62	25-Apr-95	62.91	50.78
22-Feb-95	59.84	47.89	26-Apr-95	63.81	50.45
23-Feb-95	58.36	47.44	27-Apr-95	64.96	51.26
24-Feb-95	58.76	47.75	28-Apr-95	65.33	51.19
27-Feb-95	58.97	47.19	01-May-95	64.15	51.09
28-Feb-95	57.58	46.9	02-May-95	63.65	50.95
01-Mar-95	56.74	46.44	03-May-95	62.55	50.25
02-Mar-95	55.59	46.52	04-May-95	63.59	51.27
03-Mar-95	55.94	47.41	05-May-95	63.99	51.34
06-Mar-95	56.21	46.66	08-May-95	64.21	51.15

Table 15.1 *(cont.)*

Date	Unleaded Gas	Heating Oil	Date	Unleaded Gas	Heating Oil
09-May-95	62.56	49.14	11-Jul-95	54.19	46.96
10-May-95	63.29	49.95	12-Jul-95	54.96	47.23
11-May-95	63.28	49.09	13-Jul-95	54.39	46.68
12-May-95	63.67	49.54	14-Jul-95	54.54	46.53
15-May-95	64.9	49.86	17-Jul-95	53.98	46.49
16-May-95	66.3	50.45	18-Jul-95	53.58	46.98
17-May-95	66.76	50.4	19-Jul-95	52.69	46.47
18-May-95	66.5	50.56	20-Jul-95	52.18	46.1
19-May-95	66.34	51.01	21-Jul-95	52.05	46.14
22-May-95	66.46	51.29	24-Jul-95	53.26	46.56
23-May-95	66.15	52.29	25-Jul-95	52.37	46.51
24-May-95	64.93	51.13	26-Jul-95	52.89	48.62
25-May-95	65.81	51.25	27-Jul-95	53.69	48.13
26-May-95	64.07	48.72	28-Jul-95	53.75	48
29-May-95	64.07	48.72	31-Jul-95	54.08	48.27
30-May-95	63.5	48.56	01-Aug-95	54.35	48.79
31-May-95	63	48.47	02-Aug-95	54.44	49.44
01-Jun-95	59.78	49.53	03-Aug-95	53.93	49.24
02-Jun-95	60.94	49.9	04-Aug-95	53.97	49.18
05-Jun-95	61.79	49.6	07-Aug-95	54.05	49.32
06-Jun-95	61.39	49.1	08-Aug-95	54.38	49.7
07-Jun-95	61.77	48.95	09-Aug-95	54.78	49.45
08-Jun-95	60.64	48.65	10-Aug-95	55.65	49.55
09-Jun-95	60.8	48.1	11-Aug-95	55.72	49.38
12-Jun-95	61.15	48.5	14-Aug-95	55.23	48.77
13-Jun-95	60.93	48.53	15-Aug-95	54.82	48.74
14-Jun-95	62	49.19	16-Aug-95	53.92	49.22
15-Jun-95	61.87	48.88	17-Aug-95	54.29	49.27
16-Jun-95	61.5	48.29	18-Aug-95	54.23	49.7
19-Jun-95	60.28	47	21-Aug-95	54.46	50.29
20-Jun-95	60.15	47.14	22-Aug-95	54.57	50.18
21-Jun-95	58.73	46.54	23-Aug-95	55.27	50.5
22-Jun-95	58.33	46.65	24-Aug-95	55.86	50.2
23-Jun-95	56.98	46.31	25-Aug-95	55.97	49.97
26-Jun-95	56.71	46.78	28-Aug-95	55.62	49.8
27-Jun-95	57.38	47.23	29-Aug-95	55.51	49.52
28-Jun-95	59.59	47.69	30-Aug-95	56.45	49.65
29-Jun-95	59.01	46.92	31-Aug-95	56.25	50.15
30-Jun-95	59.15	46.72	01-Sep-95	54.25	51.43
03-Jul-95	59.15	46.72	04-Sep-95	54.25	51.43
04-Jul-95	59.15	46.72	05-Sep-95	56.23	52.97
05-Jul-95	54.37	46.51	06-Sep-95	55.32	52.11
06-Jul-95	54.74	47.19	07-Sep-95	54.55	51.44
07-Jul-95	53.8	46.37	08-Sep-95	54.79	51.83
10-Jul-95	54.74	47.1	11-Sep-95	54.92	51.65

Table 15.1 *(cont.)*

Date	Unleaded Gas	Heating Oil	Date	Unleaded Gas	Heating Oil
12-Sep-95	55.74	51.95	14-Nov-95	50.43	51.56
13-Sep-95	55.34	51.25	15-Nov-95	51.24	51.71
14-Sep-95	56.81	51.8	16-Nov-95	51.55	52.22
15-Sep-95	56.63	51.53	17-Nov-95	52.79	52.96
18-Sep-95	57.73	51.65	20-Nov-95	52.9	52.73
19-Sep-95	57.23	51.37	21-Nov-95	53.12	52.28
20-Sep-95	56.39	49.3	22-Nov-95	54.12	52.54
21-Sep-95	54.87	48.67	23-Nov-95	54.12	52.54
22-Sep-95	53.49	48.09	24-Nov-95	54.12	52.54
25-Sep-95	54.01	48.85	27-Nov-95	55.45	53.42
26-Sep-95	53.79	48.23	28-Nov-95	56.24	52.95
27-Sep-95	54.55	49.02	29-Nov-95	57.45	52.2
28-Sep-95	56.05	49.5	30-Nov-95	57.36	51.62
29-Sep-95	57.67	48.65	01-Dec-95	53.02	52.67
02-Oct-95	52.78	49.26	04-Dec-95	53.56	54.03
03-Oct-95	51.93	49.28	05-Dec-95	54	54.22
04-Oct-95	50.74	48.85	06-Dec-95	53.89	54.75
05-Oct-95	48.89	47.97	07-Dec-95	54.06	55.28
06-Oct-95	49.15	48.21	08-Dec-95	54.65	56.59
09-Oct-95	50.24	48.74	11-Dec-95	54.69	56.75
10-Oct-95	50.33	48.67	12-Dec-95	55.58	56.81
11-Oct-95	50.48	48.8	13-Dec-95	57.55	57.69
12-Oct-95	49.86	48.46	14-Dec-95	57.86	57.3
13-Oct-95	50.29	48.92	15-Dec-95	59.59	57.99
16-Oct-95	50.7	48.85	18-Dec-95	59.93	59.11
17-Oct-95	50.33	48.82	19-Dec-95	59.26	59.23
18-Oct-95	49.88	48.42	20-Dec-95	57.75	59.9
19-Oct-95	49.36	48.15	21-Dec-95	56.91	60.01
20-Oct-95	49.7	48.58	22-Dec-95	57.59	60.09
23-Oct-95	49.81	48.94	25-Dec-95	57.59	60.09
24-Oct-95	49.87	49.36	26-Dec-95	58.69	60.5
25-Oct-95	49.69	49.58	27-Dec-95	60.26	62.33
26-Oct-95	50	50.44	28-Dec-95	59.28	60.32
27-Oct-95	50.06	50.34	29-Dec-95	58.6	58.63
30-Oct-95	50.74	50.59	01-Jan-96	58.6	58.63
31-Oct-95	50.83	50.4	02-Jan-96	59.09	59.93
01-Nov-95	50.55	50.95	03-Jan-96	58.74	59.44
02-Nov-95	51.72	52.04	04-Jan-96	59.44	59.28
03-Nov-95	51.51	51.72	05-Jan-96	60.48	60.64
06-Nov-95	51.03	51.15	08-Jan-96	60.48	60.64
07-Nov-95	51.14	50.99	09-Jan-96	58.65	60.43
08-Nov-95	51.42	51.45	10-Jan-96	58.19	59.59
09-Nov-95	51.06	51.62	11-Jan-96	54.44	56.16
10-Nov-95	50.7	51.63	12-Jan-96	53.1	53.57
13-Nov-95	50.3	51.57	15-Jan-96	53.9	53.3

Table 15.1 *(cont.)*

Date	Unleaded Gas	Heating Oil	Date	Unleaded Gas	Heating Oil
16-Jan-96	53.33	52.43	19-Mar-96	65.15	62.26
17-Jan-96	54.98	53.13	20-Mar-96	64.38	63.12
18-Jan-96	55.21	54.37	21-Mar-96	64.03	61.33
19-Jan-96	55.41	54.22	22-Mar-96	65.49	62.65
22-Jan-96	54.88	53.67	25-Mar-96	67	63.2
23-Jan-96	53.66	52.95	26-Mar-96	66.25	64.88
24-Jan-96	54.2	52.72	27-Mar-96	65.72	65.93
25-Jan-96	52.67	50.51	28-Mar-96	64.44	63.54
26-Jan-96	52.97	50.93	29-Mar-96	64.94	62.76
29-Jan-96	52.46	51.13	01-Apr-96	66	57.98
30-Jan-96	53.37	52.28	02-Apr-96	68.11	59.72
31-Jan-96	54.1	53.51	03-Apr-96	67.69	58.22
01-Feb-96	53.14	52.41	04-Apr-96	68.76	59.57
02-Feb-96	53.74	53.26	05-Apr-96	68.76	59.57
05-Feb-96	52.06	51.6	08-Apr-96	69.86	60.19
06-Feb-96	52.38	51.64	09-Apr-96	70.52	60.64
07-Feb-96	52.23	52.46	10-Apr-96	72.99	62.51
08-Feb-96	52.44	53.14	11-Apr-96	74.3	64.02
09-Feb-96	52.91	53.62	12-Apr-96	72.17	62.02
12-Feb-96	53	53.69	15-Apr-96	71.71	62.62
13-Feb-96	55.11	56.74	16-Apr-96	69.45	59.54
14-Feb-96	55.2	58.21	17-Apr-96	68.12	58.09
15-Feb-96	55.44	57	18-Apr-96	66.4	55.4
16-Feb-96	55.77	56.87	19-Apr-96	67.49	55.72
19-Feb-96	55.77	56.87	22-Apr-96	70.19	55.06
20-Feb-96	57.71	56.39	23-Apr-96	73.18	57.3
21-Feb-96	59.45	58.84	24-Apr-96	74.1	58.2
22-Feb-96	60.04	60.53	25-Apr-96	75.61	58.76
23-Feb-96	58.73	60.66	26-Apr-96	76.81	59.27
26-Feb-96	59.76	62.85	29-Apr-96	77.01	62.28
27-Feb-96	60.31	64.28	30-Apr-96	72.39	61.82
28-Feb-96	59.46	59.68	01-May-96	67.42	54.16
29-Feb-96	59.35	61.81	02-May-96	68.4	53.94
01-Mar-96	59.75	53.42	03-May-96	69.92	54.74
04-Mar-96	58.73	52.15	06-May-96	68.85	54.56
05-Mar-96	59.09	53	07-May-96	68.81	54.79
06-Mar-96	59.75	54.22	08-May-96	68.37	54.87
07-Mar-96	59.18	53.78	09-May-96	67.23	54.56
08-Mar-96	58.75	53.44	10-May-96	68.48	54.95
11-Mar-96	59.32	55.15	13-May-96	69.11	56.19
12-Mar-96	60.56	54.83	14-May-96	68.43	55.32
13-Mar-96	61.61	54.59	15-May-96	67.2	54.81
14-Mar-96	62.45	55.07	16-May-96	64.2	53
15-Mar-96	62.92	57.87	17-May-96	63.03	52.94
18-Mar-96	64.31	60.28	20-May-96	66.04	55.24

Table 15.1 *(cont.)*

Date	Unleaded Gas	Heating Oil	Date	Unleaded Gas	Heating Oil
21-May-96	64.95	54.06	23-Jul-96	63.08	55.94
22-May-96	64.3	54.99	24-Jul-96	61.87	55.95
23-May-96	64.25	54.39	25-Jul-96	61.66	56.25
24-May-96	64.72	54.46	26-Jul-96	60.16	55.04
27-May-96	64.72	54.46	29-Jul-96	60.52	55.19
28-May-96	63.15	54.18	30-Jul-96	61.23	55.65
29-May-96	62.36	54.06	31-Jul-96	61.8	57.08
30-May-96	59.88	52.09	01-Aug-96	61.38	57.53
31-May-96	59.12	50.85	02-Aug-96	62.12	58.71
03-Jun-96	58.99	51.25	05-Aug-96	61.31	58.29
04-Jun-96	60.69	51.52	06-Aug-96	61.23	57.43
05-Jun-96	59.39	50.85	07-Aug-96	62	58.22
06-Jun-96	60.22	51.04	08-Aug-96	62.27	58.79
07-Jun-96	60.91	51.78	09-Aug-96	61.87	58.49
10-Jun-96	61.4	51.4	12-Aug-96	62.89	59.56
11-Jun-96	60.8	50.79	13-Aug-96	63.09	60.01
12-Jun-96	59.68	50.88	14-Aug-96	62.49	60.41
13-Jun-96	58.89	50.95	15-Aug-96	61.96	59.68
14-Jun-96	59.5	51.55	16-Aug-96	63.38	61.63
17-Jun-96	61.21	53.34	19-Aug-96	65.27	62.58
18-Jun-96	60.24	52.5	20-Aug-96	64.01	61.67
19-Jun-96	57.96	51.12	21-Aug-96	63.12	60.98
20-Jun-96	58.68	51.53	22-Aug-96	63.88	62.48
21-Jun-96	58.74	51.36	23-Aug-96	63.22	61.99
24-Jun-96	58.23	51.3	26-Aug-96	61.62	61.03
25-Jun-96	57.46	51.17	27-Aug-96	61.21	61.13
26-Jun-96	58.36	52.34	28-Aug-96	62.33	62.04
27-Jun-96	59.36	53.64	29-Aug-96	63.72	63.67
28-Jun-96	60.03	53.95	30-Aug-96	62.82	62.82
01-Jul-96	61.51	55.14	02-Sep-96	62.82	62.82
02-Jul-96	60.89	54.28	03-Sep-96	62.96	65.07
03-Jul-96	62.47	54.71	04-Sep-96	62.96	64.21
04-Jul-96	62.47	54.71	05-Sep-96	64.41	65.03
05-Jul-96	62.47	54.71	06-Sep-96	65.27	66.4
08-Jul-96	61.68	54.89	09-Sep-96	64.09	65.95
09-Jul-96	61.81	55.26	10-Sep-96	64.85	66.67
10-Jul-96	63.11	55.59	11-Sep-96	65.91	68.19
11-Jul-96	64.59	56.7	12-Sep-96	65.91	69.17
12-Jul-96	64	56.62	13-Sep-96	64.6	67.94
15-Jul-96	65.56	57.72	16-Sep-96	62.87	65.29
16-Jul-96	65.13	57.18	17-Sep-96	62.74	65.59
17-Jul-96	63.89	56.32	18-Sep-96	63.06	67.87
18-Jul-96	63.87	56.74	19-Sep-96	61.32	66.77
19-Jul-96	62.41	56.02	20-Sep-96	61.09	67.42
22-Jul-96	62.76	55.85	23-Sep-96	60.07	67.48

Table 15.1 *(cont.)*

Date	Unleaded Gas	Heating Oil	Date	Unleaded Gas	Heating Oil
24-Sep-96	62.83	69.69	26-Nov-96	69.01	71.24
25-Sep-96	63.1	71.77	27-Nov-96	69.35	71.97
26-Sep-96	62.99	70.9	28-Nov-96	69.35	71.97
27-Sep-96	64.6	71.49	29-Nov-96	69.35	71.97
30-Sep-96	62.71	71.51	02-Dec-96	68.12	73.57
01-Oct-96	62.82	70.76	03-Dec-96	69.13	74.22
02-Oct-96	62.42	71.98	04-Dec-96	68.24	73.57
03-Oct-96	63.68	74.69	05-Dec-96	69.68	75.11
04-Oct-96	63.63	74.43	06-Dec-96	69.8	74.66
07-Oct-96	66.34	76.49	09-Dec-96	68.88	72.13
08-Oct-96	66.5	76.19	10-Dec-96	66.86	69.62
09-Oct-96	65.59	73.97	11-Dec-96	63.56	66.82
10-Oct-96	63.52	70.92	12-Dec-96	64.72	68.67
11-Oct-96	65.52	71.43	13-Dec-96	67.04	71.71
14-Oct-96	67.7	74.07	16-Dec-96	69.52	74.82
15-Oct-96	67.08	73.07	17-Dec-96	69.77	73.54
16-Oct-96	65.45	71.56	18-Dec-96	71.17	74.18
17-Oct-96	66.53	72.29	19-Dec-96	71.22	73.78
18-Oct-96	67.94	74.06	20-Dec-96	70.19	72.97
21-Oct-96	67.92	73.63	23-Dec-96	68.9	71.08
22-Oct-96	69.12	73.45	24-Dec-96	69.56	71.4
23-Oct-96	68.16	70.96	25-Dec-96	69.56	71.4
24-Oct-96	69.22	70.49	26-Dec-96	69.51	70.06
25-Oct-96	70.1	71.72	27-Dec-96	69.74	70.55
28-Oct-96	70.3	71.46	30-Dec-96	69.61	70.57
29-Oct-96	69.1	69.83	31-Dec-96	70.67	72.84
30-Oct-96	70	68.46	01-Jan-97	70.67	72.84
31-Oct-96	66.56	66.34	02-Jan-97	71.1	72.11
01-Nov-96	64.7	66.6	03-Jan-97	70.7	71.29
04-Nov-96	65	65.95	06-Jan-97	72.52	73.64
05-Nov-96	64.61	65.42	07-Jan-97	72.1	72.49
06-Nov-96	63.63	66.45	08-Jan-97	72.19	73.43
07-Nov-96	63.8	66.89	09-Jan-97	70.48	73.05
08-Nov-96	65.27	68.93	10-Jan-97	70.36	72.15
11-Nov-96	65.02	68.35	13-Jan-97	68.09	69.7
12-Nov-96	65.77	68.25	14-Jan-97	67.04	69.42
13-Nov-96	68.34	71.2	15-Jan-97	68.85	71.42
14-Nov-96	68.92	73.4	16-Jan-97	68.69	69.92
15-Nov-96	66.92	72.61	17-Jan-97	68.09	68.44
18-Nov-96	65.77	71.85	20-Jan-97	67.23	66.94
19-Nov-96	67.39	73.68	21-Jan-97	67.44	66.03
20-Nov-96	65.39	72.09	22-Jan-97	68.22	66.89
21-Nov-96	67.04	73.85	23-Jan-97	68.42	66.35
22-Nov-96	67.8	72.79	24-Jan-97	67.75	66.77
25-Nov-96	67.99	72.23	27-Jan-97	67.62	67.29

Table 15.1 *(cont.)*

Date	Unleaded Gas	Heating Oil	Date	Unleaded Gas	Heating Oil
28-Jan-97	67.04	66.83	01-Apr-97	62.67	53.95
29-Jan-97	68.23	68.84	02-Apr-97	60.61	52.52
30-Jan-97	69.82	70.34	03-Apr-97	60.9	53.26
31-Jan-97	68.47	68.65	04-Apr-97	60.48	53.14
03-Feb-97	68.35	65.28	07-Apr-97	60.72	53.13
04-Feb-97	68.31	64.18	08-Apr-97	61.17	52.89
05-Feb-97	67.54	63.32	09-Apr-97	60.7	53.11
06-Feb-97	65.3	61.45	10-Apr-97	61.07	54.86
07-Feb-97	63.06	60.53	11-Apr-97	60.88	53.87
10-Feb-97	63.53	61.76	14-Apr-97	61.96	54.67
11-Feb-97	63.96	61.86	15-Apr-97	61.9	54.85
12-Feb-97	62.89	60.85	16-Apr-97	60.38	53.48
13-Feb-97	63.18	59.92	17-Apr-97	60.7	54
14-Feb-97	64.25	60.81	18-Apr-97	61.49	54.68
17-Feb-97	64.25	60.81	21-Apr-97	62.8	55.48
18-Feb-97	64.16	59.42	22-Apr-97	61.77	54.83
19-Feb-97	64.68	59.59	23-Apr-97	61.74	55.65
20-Feb-97	62.78	58.04	24-Apr-97	62.84	55.89
21-Feb-97	61.82	57.85	25-Apr-97	62.5	55.9
24-Feb-97	60.24	55.47	28-Apr-97	62.34	56.53
25-Feb-97	62.23	56.82	29-Apr-97	63.36	58.91
26-Feb-97	62.26	56.68	30-Apr-97	63.91	58.07
27-Feb-97	62.67	56.03	01-May-97	62.63	54.33
28-Feb-97	61.65	54.76	02-May-97	60.52	53.02
03-Mar-97	61.77	53.18	05-May-97	60.54	53.05
04-Mar-97	62.89	53.34	06-May-97	60.31	53.53
05-Mar-97	63.33	52.54	07-May-97	60.92	53.08
06-Mar-97	64.48	53.43	08-May-97	62.5	54.38
07-Mar-97	65.67	54.08	09-May-97	62.89	54.52
10-Mar-97	64.36	53.08	12-May-97	64.47	56.65
11-Mar-97	63.86	52.83	13-May-97	64.76	56.48
12-Mar-97	64.63	54.08	14-May-97	64.38	56.42
13-Mar-97	64.23	54.22	15-May-97	64.04	56.48
14-Mar-97	65.77	55.33	16-May-97	65.87	58.47
17-Mar-97	65.26	54.3	19-May-97	65.21	57.92
18-Mar-97	67.48	56.18	20-May-97	65.39	57.64
19-Mar-97	67.96	56.29	21-May-97	66.53	57.55
20-Mar-97	67.58	55.94	22-May-97	66.93	57.8
21-Mar-97	67.64	55.98	23-May-97	66.92	57.52
24-Mar-97	66.51	55.73	26-May-97	66.92	57.52
25-Mar-97	66.52	56.83	27-May-97	65.38	55.27
26-Mar-97	64.82	55.43	28-May-97	65.8	55.39
27-Mar-97	64.63	56.07	29-May-97	65.15	56
28-Mar-97	64.63	56.07	30-May-97	63.68	56.49
31-Mar-97	63.68	56.72	02-Jun-97	63.68	56.32

Table 15.1 *(cont.)*

Date	Unleaded Gas	Heating Oil	Date	Unleaded Gas	Heating Oil
03-Jun-97	61.44	54.62	05-Aug-97	67.1	58.32
04-Jun-97	60.42	54.16	06-Aug-97	66.06	56.98
05-Jun-97	59.82	53.32	07-Aug-97	64.33	55.3
06-Jun-97	57.13	51.52	08-Aug-97	61.99	54.29
09-Jun-97	56.2	51.5	11-Aug-97	61.47	54.36
10-Jun-97	56.4	51.65	12-Aug-97	63.71	55.1
11-Jun-97	56.54	51.52	13-Aug-97	66.08	56.04
12-Jun-97	57.08	51.62	14-Aug-97	66.33	55.87
13-Jun-97	57.4	51.64	15-Aug-97	66.81	55.25
16-Jun-97	58.03	51.94	18-Aug-97	65.44	55.09
17-Jun-97	58.48	52.45	19-Aug-97	67.58	55.71
18-Jun-97	56.78	51.44	20-Aug-97	69.64	55.1
19-Jun-97	56.09	51.45	21-Aug-97	67.15	53.48
20-Jun-97	55.48	51.33	22-Aug-97	67.48	53.41
23-Jun-97	55.64	51.92	25-Aug-97	64.5	52.2
24-Jun-97	55.68	51.57	26-Aug-97	63.81	52.09
25-Jun-97	56.97	52.99	27-Aug-97	66.4	53.26
26-Jun-97	56.79	52.02	28-Aug-97	67.51	52.51
27-Jun-97	57.91	53.33	29-Aug-97	68.82	51.85
30-Jun-97	58.12	53.7	01-Sep-97	68.82	51.85
01-Jul-97	58.78	54.84	02-Sep-97	62.79	53.4
02-Jul-97	59.29	54.92	03-Sep-97	62.55	53.35
03-Jul-97	57.92	52.76	04-Sep-97	59.92	52.54
04-Jul-97	57.92	52.76	05-Sep-97	60.12	53.78
07-Jul-97	57.94	52.78	08-Sep-97	59.32	53.14
08-Jul-97	58.92	53	09-Sep-97	59.49	52.83
09-Jul-97	58.23	52.65	10-Sep-97	58.33	51.57
10-Jul-97	58.6	52.11	11-Sep-97	58.78	52.05
11-Jul-97	59.26	52.35	12-Sep-97	58.77	52.58
14-Jul-97	58.35	51.67	15-Sep-97	58.22	52.52
15-Jul-97	59.94	52.95	16-Sep-97	59.04	53.85
16-Jul-97	60.46	52.68	17-Sep-97	58.45	53.35
17-Jul-97	61.89	53.89	18-Sep-97	57.25	53.44
18-Jul-97	60.05	52.22	19-Sep-97	57.48	53.45
21-Jul-97	60.04	52.35	22-Sep-97	58.58	54.73
22-Jul-97	60.02	52.7	23-Sep-97	58.36	54.64
23-Jul-97	61.12	53.28	24-Sep-97	58.37	55.24
24-Jul-97	62.21	53.39	25-Sep-97	59.25	56.51
25-Jul-97	64.03	53.99	26-Sep-97	61.34	57.92
28-Jul-97	64.84	54.14	29-Sep-97	63.13	59.25
29-Jul-97	66.47	54.31	30-Sep-97	62.63	58.77
30-Jul-97	69.9	55.78	01-Oct-97	59.9	58.19
31-Jul-97	67.84	55.61	02-Oct-97	61.43	59.8
01-Aug-97	65.07	56.56	03-Oct-97	62.99	62.01
04-Aug-97	66.74	58.44	06-Oct-97	61.3	59.69

Table 15.1 *(cont.)*

Date	Unleaded Gas	Heating Oil	Date	Unleaded Gas	Heating Oil
07-Oct-97	60.91	59.6	07-Nov-97	59.95	57.99
08-Oct-97	61.5	60.16	10-Nov-97	59.18	57.28
09-Oct-97	61.18	60.08	11-Nov-97	59.06	57.82
10-Oct-97	61.24	59.95	12-Nov-97	58.62	57.92
13-Oct-97	59.83	58.27	13-Nov-97	59.55	58.62
14-Oct-97	58.88	57.01	14-Nov-97	60.99	59.54
15-Oct-97	58.2	56.94	17-Nov-97	59.44	57.85
16-Oct-97	59.68	58.01	18-Nov-97	58.65	57.61
17-Oct-97	59.31	57.4	19-Nov-97	58.65	56.67
20-Oct-97	59.66	57.82	20-Nov-97	57.22	55.45
21-Oct-97	59.08	57.64	21-Nov-97	57.74	55.48
22-Oct-97	60.79	58.77	24-Nov-97	58.69	55.6
23-Oct-97	60.26	58.09	25-Nov-97	59.08	55.49
24-Oct-97	59.6	57.03	26-Nov-97	57.31	53.1
27-Oct-97	59.95	57.74	27-Nov-97	57.31	53.1
28-Oct-97	58.88	56.52	28-Nov-97	57.31	53.1
29-Oct-97	60.09	57.19	01-Dec-97	56.25	52.71
30-Oct-97	60.67	58.12	02-Dec-97	56.43	53.25
31-Oct-97	60.22	57.77	03-Dec-97	56.55	53.5
03-Nov-97	59.8	58.78	04-Dec-97	56.34	53.35
04-Nov-97	58.96	58.11	05-Dec-97	56.59	53.38
05-Nov-97	58.2	57.18	08-Dec-97	56.96	53.52
06-Nov-97	59.26	57.43			

Index

Printed in the United States
By Bookmasters